Chemical Engineers in Medicine 2020

Topical Conference at the 2020 AIChE Annual Meeting

Online
16 – 20 November 2020

ISBN: 978-1-7138-2288-2

AA009126

Printed from e-media with permission by:

Curran Associates, Inc.
57 Morehouse Lane
Red Hook, NY 12571

Some format issues inherent in the e-media version may also appear in this print version.

Copyright© (2020) by AIChE
All rights reserved.

Printed with permission by Curran Associates, Inc. (2021)

For permission requests, please contact AIChE
at the address below.

AIChE
120 Wall Street, FL 23
New York, NY 10005-4020

Phone: (800) 242-4363
Fax: (203) 775-5177

www.aiche.org

Additional copies of this publication are available from:

Curran Associates, Inc.
57 Morehouse Lane
Red Hook, NY 12571 USA
Phone: 845-758-0400
Fax: 845-758-2633
Email: curran@proceedings.com
Web: www.proceedings.com

TABLE OF CONTENTS

600369 Development of an Electroactive Platform for Detection of Virus Fusion to Host Membranes

Abstract:

Every year, new virus outbreaks are reported and have potential to cause pandemics with devastating socio-economic impacts, with the coronavirus disease 19 (COVID-19) serving as a prime example. Recent viral outbreaks in this decade have alluded to the importance of quick and accurate viral sensing strategies so that the infected individual can begin their treatment and limit contact with uninfected individuals quicker, mitigating the impact of an outbreak. PCR, one of the most common viral detection methods, is a specific and sensitive detection technique, but requires trained personnel and sophisticated machinery, which can slow response times. An ideal detection method should integrate the specificity and sensitivity with the speed necessary to diagnosis patients rapidly.

Enveloped viruses (such as influenza virus, coronavirus) enter host cells by binding to a receptor and then fusing its viral membrane to the host cell membrane to insert its genome. We propose a novel sensing platform that leverages the virus's propensity to bind specifically and fuse to a host cell membrane. We can couple a biomimetic membrane on top of an electrically conductive polymer (PEDOT:PSS) to probe the electrical responses of the system once viral fusion has occurred. Successful viral fusion to the membrane in a matter of minutes may cause a detectable change in electrical response, serving as the readout of a viral infection. Using influenza virus as a proof-of-principle, we first show that we can form the biomimetic membrane incorporating the influenza viral receptor on PEDOT:PSS. We then visualize influenza single virion fusion on PEDOT:PSS and demonstrate that the kinetics of this fusion event are comparable to influenza fusion on a non-conductive polymer. Lastly, using electrical impedance spectroscopy, we provide evidence that influenza viral fusion causes a detectable change in electrical response, demonstrating that it is possible to exploit viral membrane fusion as a diagnostic strategy.

Copyright © American Institute of Chemical Engineers. All rights reserved.

599943 Engineering DNA-Based Materials for the Analysis of Live Single Cells

Abstract:

The ability to probe the composition of live cells is important for learning about chemical processes inside of cells as well as for creating diagnostic platforms based on molecular profiling. However, current strategies for detecting analytes in single living cells are severely limited both in terms of probe design and sensing strategy. While fluorescent proteins and nucleic acid-based aptamers have revolutionized live-cell imaging, these sensors must be genetically encoded into the cells and therefore have limited applications in a clinical setting. Many techniques require cell lysis or a bulk population of cells to yield a detectable quantity of the target analyte. Additionally, a major impediment to the development of intracellular probes is the inability of most probes to traverse the cell membrane. Consequently, these probes require the use of transfection reagents or physical methods that perturb the cell membrane, which in turn can interfere with the measurements by changing the analyte's expression profiles. Furthermore, existing strategies for signal transduction to indicate the presence of a given analyte, such as fluorophore-quencher-based systems and FRET-based methodologies, are limited and susceptible to false-positive signal. Therefore, the development of a versatile platform for intracellular analyte detection in live cells is an outstanding challenge.

To address this challenge, I have first developed DNA-based aptamers that utilize a novel sensing strategy. Specifically, these aptamers are chemically modified at a single base with a viscosity-sensitive dye. Target binding to the aptamer changes the conformation of the aptamer and forces the dye to intercalate (or FIT) between the DNA base pairs, turning its fluorescence on. This "FIT" strategy avoids false-positive signals common to most fluorophore-quencher-based systems, quantitatively outperforms FRET-based techniques, gives up to 20-fold fluorescence enhancement in buffer, and allows for the detection of analytes in complex biological milieu such as human serum.

To apply this technique to the analysis of intracellular analytes, I have then developed a new platform based on protein spherical nucleic acids (ProSNAs). ProSNAs are composed of a protein core surrounded by a DNA shell. The dense functionalization and radial arrangement of the DNA around the protein core allows efficient cellular uptake of the ProSNA without the use of transfection reagents or physical stimuli. By using a FIT-aptamer as the DNA sequence, this design allows for the detection of intracellular analytes. As an example, I demonstrate that using an i-motif sequence that undergoes pH-dependent conformational changes allows for the detection of intracellular pH. As more than 500 aptamers are known for various targets, our nano-construct is a powerful platform for the detection of intracellular analytes. Moreover, by using a functional protein as the core, it is possible to detect analytes for which aptamers do not exist and further expand the range of analytes that can be detected. Using glucose oxidase as the core, I show that glucose can be detected (for which is there is no known aptamer with a biologically relevant binding affinity) in 8 different cell lines. Taken together, these constructs represent a plug-and-play platform for detecting a wide variety of intracellular analytes with great potential in the field of biodiagnostics.

References

†Equal Contribution

Samanta, D.†; **Ebrahimi, S. B.**†; Mirkin, C. A., "*Nucleic-Acid Structures as Intracellular Probes for Live Cells,*" *Advanced Materials* **2020**, *32*, 1901743.

Ebrahimi, S. B.†; Samanta, D.†; Cheng, H. F.; Nathan, L. I.; Mirkin, C. A., "*Forced Intercalation (FIT)-Aptamers,*" *Journal of the American Chemical Society* **2019**, *141*, 13744.

Samanta, D.; Iscen, A.; Laramy, C. R.; **Ebrahimi, S. B.**; Bujold, K. E.; Schatz, G. C.; Mirkin, C. A., "*Multivalent Cation-Induced Actuation of DNA-Mediated Colloidal Superlattices,*" *Journal of the American Chemical Society* **2019**, *141*, 19973.

Copyright © American Institute of Chemical Engineers. All rights reserved.

608845 Molecular Engineering of a Colorimetric Nanogel Sensor for Clinical Radiotherapy and Trauma Monitoring

Abstract:

Radiotherapy using ionizing radiation (e.g. X rays) remains a mainstay of treatment modalities for different types of cancers. Overexposure to radiation in the clinic or as part of an adverse event can induce tissue toxicity and damage, whereas underdosing can lead to poor efficacies of tumor ablation. Effective radiation sensors are critical in ensuring optimum delivery to patients undergoing radiotherapy and for patient stratification in cases of radiation exposure. Currently existing sensors possess several disadvantages in rapid radiation detection because of their fragility, sensitivity to light and heat, dose-rate dependence, long processing times, lack of conformation to tissue morphologies and / or high costs. Commonly used Nanodots™ are expensive, require separate read-out devices and do not conform to tissue morphologies. We developed a novel colorimetric nanogel sensor, based on the formation of maroon-colored gold nanoparticles from their colorless precursor salt formulations, upon exposure to different levels and types of ionizing radiation. By measuring the intensity of color developed with a simple bench-top absorbance spectrophotometer we were able to detect doses typically administered in fractionated cancer radiotherapy (i.e. ranging from 0-5 Gy doses) and beyond. Topographical distribution of delivered radiation doses was qualitatively and quantitatively determined following addition of quenching agents, and the efficacy of the nanogel sensor was demonstrated for different sources of ionizing radiation including protons, electrons, photons and radioactive isotopes. The translational potential of the sensor was demonstrated using anthropomorphic phantoms and in live canine patients undergoing clinical radiotherapy treatments. Our results demonstrate a new generation of nanogel sensors for colorimetric detection of radiation for improving patient safety in clinical radiotherapy and trauma care.

Copyright © American Institute of Chemical Engineers. All rights reserved.

609804 Unique Electrochemical Detection of Sepsis Using Triplex Biomarker Detection Panel with IL-6, IL-8 and IL-10 in Blood Plasma

Abstract:

Sepsis is a life-threatening disease that is caused by the body's extreme reaction to an infection which could lead to tissue damage, multiple organ failure or even death. It is one of the leading causes of death worldwide, estimated to affect more than 30 million people every year according to the world health organization. Despite scientific developments in health care technologies in the past three decades, a substantial challenge continues to persist for early rapid diagnosis of sepsis. Thorough and timely diagnosis plays a substantial role in increasing the survival chances for patients with adequate treatment, thus reducing mortality. However, current gold standard blood culture techniques require large patient sample volumes and provides delayed results (6 hours to up to 5 days). This may lead to clinical ambiguity in treatment measures, propelling the body's immune response to spiral uncontrollably into septic shock or severe sepsis. This fails to deliver the required speed and accuracy to enable accurate diagnosis. To overcome existing challenges, we tailored a specific biosensor to screen for sepsis as a rapid point-of-care (POC) technology enabling multiplexed detection capabilities. The novelty presented in this research combines the effectiveness of specific multiplexed biomarkers for detection of sepsis with electrochemical impedance spectroscopy (EIS) technique to enable clinically efficient decision-making process. This work leverages the detection of pro and anti-inflammatory cytokine biomarkers such as interleukin 6 (IL-6), interleukin-8 (IL-8) and interleukin-10 (IL-10) in blood plasma. Researchers suggests good correlation between timing of pro and inflammatory cytokine release to degree of sepsis, associated with deleterious effects. The developed label-free, point-of-care biosensor is surface engineered with specific capture probes that utilizes EIS to measure the capacitive impedance change reflecting binding interactions between the capture probe and target biomarker enabling simultaneous triplex detection. Specificity of the POC biosensor was validated using cross-reactive studies, which displayed insignificant interference from non-specific biomarkers. The biosensor also displays stable and repeatable performance. Key advantage of the triplex biosensor is the use of undiluted blood plasma volume of less than 40 μL, providing result within 15 minutes. Additionally, the triplex biosensor technology was validated with 20 patient samples that were clinically tested positive for sepsis. This unique sensor technology would allow medical practitioners to facilitate targeted interventions for septic patients as a rapid prognostic approach, preventing complications arriving from sepsis.

Copyright © American Institute of Chemical Engineers. All rights reserved.

598873 Harnessing the Protein Corona Towards Carbon Nanotube-Based Sensor Design

Abstract:

Unpredictable protein adsorption on nanoparticles remains a considerable challenge towards effectively applying nanotechnologies, such as nanosensors, in biological environments. However, if well understood and controlled, proteins adsorbed on nanoparticles can be exploited to develop protein-based nanosensors.

Herein, we employ our knowledge of spontaneous "protein corona" formation on single-walled carbon nanotubes (SWCNTs) in biofluids towards rational nanosensor design based on protein-SWCNT complexation.[1] SWCNTs are implemented as the sensor platform due to their desirable optical and physical characteristics, including photostable fluorescence in the near-infrared tissue transparency window[2] and high aspect ratio that confers delivery capabilities.[3] We first study the protein corona composition formed on SWCNT-based probes within relevant biofluids such as blood plasma and cerebrospinal fluid.[4] Then, we take advantage of the intrinsic binding affinity of specific proteins on the carbon nanotube to selectively adsorb proteins that, in turn, bind to their natural ligand partners. Thus, the adsorbed protein acts as the recognition moiety that interacts with the environment and detects the analyte of interest, tethered to the SWCNT that serves as the signal transducer. This nanosensor platform combines the sensitive and selective binding capabilities of proteins for their ligand partners with the near-infrared fluorescence readout of SWCNTs that is modulated upon protein-ligand binding. In this manner, we demonstrate that although formation of the nanoparticle protein corona can impede the intended nanobiotechnology application, it also presents an opportunity to create improved protein-nanoparticle architectures by manipulating selective protein adsorption on the nanoparticle surface.

References

1. Pinals, R. L., Yang, D., Lui, A., Cao, W. & Landry, M. P. Corona Exchange Dynamics on Carbon Nanotubes by Multiplexed Fluorescence Monitoring. *J. Am. Chem. Soc.* (2019) doi:10.1021/jacs.9b09617.
2. Boghossian, A. A. *et al.* Near-Infrared Fluorescent Sensors based on Single-Walled Carbon Nanotubes for Life Sciences Applications. *ChemSusChem* **4**, 848–863 (2011).
3. Demirer, G. S. *et al.* High aspect ratio nanomaterials enable delivery of functional genetic material without DNA integration in mature plants. *Nat. Nanotechnol.* **14**, 456–464 (2019).
4. Pinals, R. L. *et al.* Protein Corona Composition and Dynamics on Carbon Nanotubes in Blood Plasma and Cerebrospinal Fluid. *bioRxiv* 2020.01.13.905356 (2020) doi:10.1101/2020.01.13.905356.

Copyright © American Institute of Chemical Engineers. All rights reserved.

607484 Point-of-Care Cancer Biomarker Detection System Integrating Surface Acoustic Wave Streaming and Metal-Enhanced Fluorescence

Abstract:

Point-of-care cancer biomarker detection system integrating surface acoustic wave streaming and metal-enhanced fluorescence

Yuqi Huang, Shuangming Li and Venkat R. Bhethanabotla

yuqi@usf.edu

shuangming@mail.usf.edu

bhethana@usf.edu

Department of Chemical and Biomedical Engineering, University of South Florida, Tampa, Florida 33620-5350, USA

Abstract:

Immunofluorescence is one of the primary techniques for microbiological assays. It can help to visualize, identify, and even quantify the interaction between antibody and antigen, some even used for DNA and protein detection.[1,2] However, there is a clinical need of increasing its limit of detection, specificity and portability. In this work, a portable optical fluorescence system was designed to be combined with a direct digital synthesizer for surface acoustic wave (SAW) generation and metal-enhanced fluorescence (MEF) to lower the detection limit in immunofluorescence assays.[3,4] MEF was achieved by depositing silver nanoparticles with covered silica film, which can plasmonically enhance fluorescence intensity. Rayleigh wave streaming was generated on a piezoelectric device for surface acoustic wave streaming purpose, which can prevent nonspecific binding on the biological interface.[5] The miniaturized prototype, integrating acoustic wave removal of interfering proteins combined with optical detection, was developed to realize immune detection on a portable platform with limit of detection at nanogram per milliliter level in detecting carcinoembryonic antigen (CEA) from human plasma.

Recent work shows that the MEF substrate fabricated with the above method can lower the detection limit of CEA from PBS to 100pg/mL level. However, testing the detection limit for CEA from human plasma has raised to 500ng/mL. Non-specific proteins, dusts and electrolytes in human plasma are primary suspects to raise background noise which increases the limit of detection. Further tests show that by adding SAW streaming to the sample tested on MEF substrate, the detection limit brings back down to 50pg/mL. A reasonable inference to this detection limit drop is the removal of non-specific protein by streaming surface acoustic wave to the sample surface, where the non-specific absorbents are removed by shear force from the wave, and the background noise is reduced to lower the limit of detection for CEA from human plasma.

1. K. Aslan, I. Gryczynski, J. Malicka, E. Matveeva, J. R. Lakowicz, and C. D. Geddes, "Metal-enhanced fluorescence: an emerging tool in biotechnology," Current Opinion in Biotechnology, vol. 16, pp. 55-62, **2005**.
2. R. A. Goldsby, T. J. Kindt, B. A. Osborne, and J. Kuby, "Immunology," ed: New York: WH Freeman, **2003**.
3. S. Li and V. Bhethanabotla, "Design of a Portable Orthogonal Surface Acoustic Wave Sensor System for Simultaneous Sensing, Removal of Nonspecifically bound Proteins and Mixing", Sensors, vol. 19(18), **2019**.
4. Liu, J.; Li, S.; Bhethanabotla, V. R., Integrating Metal-Enhanced Fluorescence and Surface Acoustic Waves for Sensitive and Rapid Quantification of Cancer Biomarkers from Real Matrices. *ACS Sensors*, *3* (1), 222-229, **2018**.
5. S. Cular, D. W. Branch, V. R. Bhethanabotla, G. D. Meyer, and H. G. Craighead, "Removal of nonspecifically bound proteins on microarrays using surface acoustic waves," IEEE Sensors journal, vol. 8, pp. 314-320, **2008**.

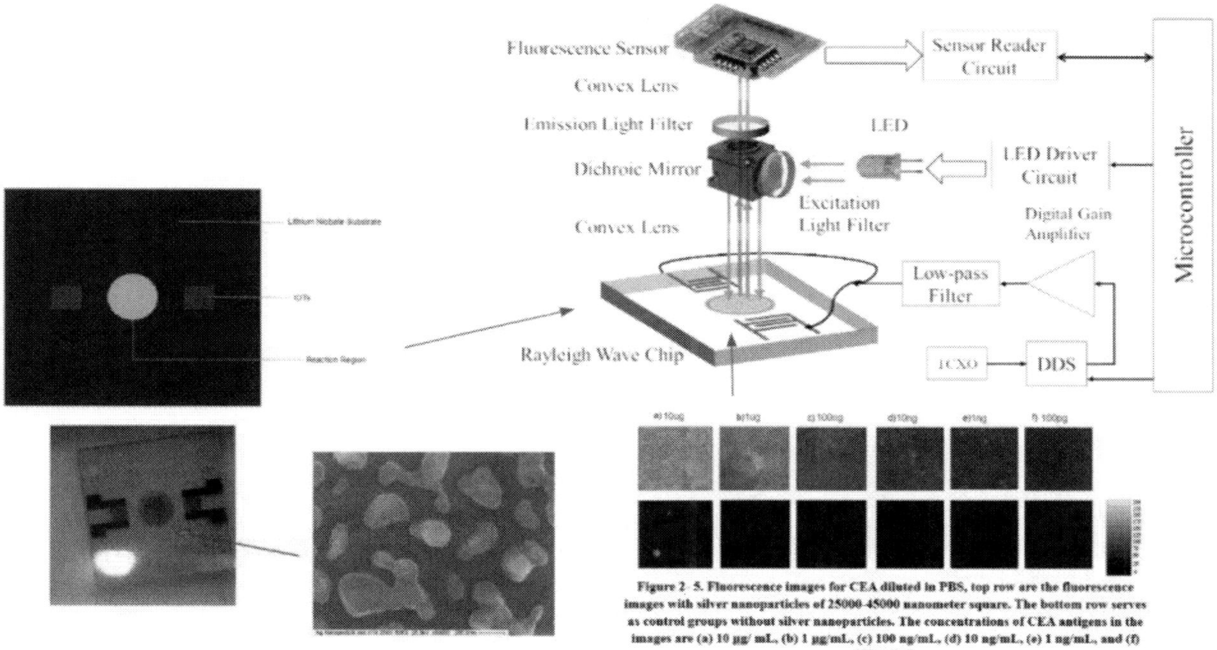

Figure 2-5. Fluorescence images for CEA diluted in PBS, top row are the fluorescence images with silver nanoparticles of 25000-45000 nanometer square. The bottom row serves as control groups without silver nanoparticles. The concentrations of CEA antigens in the images are (a) 10 μg/ mL, (b) 1 μg/mL, (c) 100 ng/mL, (d) 10 ng/mL, (e) 1 ng/mL, and (f) 100 pg/mL.

Copyright © American Institute of Chemical Engineers. All rights reserved.

609302 High-Throughput Quantification of Influenza A Virus RNA Using Novel Drop-Based qRT-PCR Analysis

Abstract:

Influenza A virus (IAV) causes seasonal epidemics and sporadic worldwide pandemics that have resulted in over 100 million deaths during the past century. IAV populations are dynamic with high mutation rates and short replication times. A better understanding of IAV diversity at the single cell level would provide greater insight into how heterogeneity of infection affects population-level dynamics. Investigating IAV and host cell dynamics is typically performed using bulk culture and may not fully capture cell-to-cell differences in virus production. Thus, there is a need for high-throughput quantification of the IAV output, or burst size, from single cells to elucidate individual host-cell dynamics at the population scale. Here, we have developed a novel drop-based quantitative reverse transcriptase polymerase chain reaction (qRT-PCR) method to measure the amount of IAV RNA contained in individual drops. Concentrations of RNA are encapsulated in tens of thousands of picoliter-sized drops. These drops are processed on a microfluidic device which splits a small portion of the drop and merges it with qRT-PCR solution. The merged drops are thermocycled using a standard qPCR machine. Drops are randomly sampled at different cycles to measure drop fluorescence. Using a novel qPCR analysis method, we quantify the amount of RNA contained in the drops. Our method is the first to measure IAV RNA using a high-throughput drop-based assay, which will allow for future work in investigating the heterogeneity of IAV infection.

Copyright © American Institute of Chemical Engineers. All rights reserved.

597969 Microfluidic Paper-Based Analytical Devices Using Plasma Processes

Abstract:

Disability adjusted life years (DALY) are the sum of years of life lost due to premature mortality and years lived with a disability. In 2010, DALY were approximately 2.4 billion. It is estimated that ~90% of DALY are due to communicable and non-communicable diseases. This problem is even more severe in developing countries where the major stumbling block for disease control is lack of proper disease diagnostic methods. Point-of care testing (POTC) methods in which users do not require specialized equipment and sophisticated lab facilities for disease diagnostics will be ideal for developing countries. Therefore, the World Health Organization (WHO) has stipulated a set of criteria for POCT in resource-limited settings called ASSURED criteria—affordability, sensitivity, specificity, user-friendly, rapid and robust, equipment-free and deliverable to those who need it. Due to the necessity and increasing demand of personalized medicine, the global POTC market is expected to grow from US \$23.16 billion in 2016 to US \$36.96 billion in 2021 at the compound annual growth rate of 9.8% from 2016 to 2021.

Microfluidic paper-based analytical devices (μ-PAD) have shown considerable promise to meet the ASSURED criteria set by the WHO. Paper has been patterned using various fabrication methods to allow distribution of liquid samples to targeted locations where subsequent reactions with reagents deposited in the paper are performed. These reactions develop color signals that can then be read using a smartphone. Low cost, portability, and the absence of external power source requirements make these devices desirable for POTC in less developed areas. It has been more than a decade since the first μ-PAD was invented. Yet these μ-PADs are not yet commercialized because of multiple technical problems including 1) contamination and evaporation of samples because of the lack of proper μ-PAD packaging, 2) inability to carry out complex tests that require multiple reagent additions, and 3) loss of sample in flow channels.

To address the issue of contamination and evaporation, a fully enclosed paper microfluidic device has been fabricated using pentafluoroethane (PFE) plasma deposition followed by O_2 plasma etching. Structures with one and two non-interacting, fully enclosed hydrophilic channels were generated in a single paper sheet using metal masks. Furthermore, by performing an additional O_2 plasma step with a secondary mask, pinholes were created at the reaction zones for reagent loading. Finally, to demonstrate the functionality of the device, a glucose assay was performed. Quantitative analysis of glucose assays showed that the device can be used for clinically relevant range of glucose. To our knowledge, this is the first time that such structures have been fabricated without paper stacking. Multi-layer devices have enhanced functionality relative to a single channel device, because the design space for creating networks of channels within the paper substrate is greatly expanded. The fluid-filled channels in the fabricated device are isolated, thereby preventing contamination due to handling and environmental exposure. Fluid evaporation can be inhibited by sealing the device with adhesive tape without contaminating the enclosed channels hence preventing loss of sample due to evaporation; a critical factor while working with precious samples.

To address the challenge of fluid flow control in μ-PADs, we extended our work on fully enclosed channel fabrication inside paper substrates to incorporate flow control functionality into the channels. Using pentafluoroethane plasma deposition and subsequent O_2 plasma etching with metal masks, fully enclosed hydrophilic channels with varying wettability were designed and fabricated to precisely control fluid flow in μ-PADs. Flow control is one of the most critical functionalities in μ-PADs because it allows controlled implementation of multiple sequential reactions on a single device with minimal user interference. Potential benefits of flow control include improved limit of detection, ability to carry out complex diagnostic tests, reduced assay time, and decreased likelihood of human error. We studied the effect of controlled fluorocarbon removal on the surface of cellulose fibers on channel wettability and demonstrated that this behavior can be exploited to significantly vary the fluid velocity in the flow channels. Furthermore, by simply covering the flow channels with adhesive tape, we demonstrated that the effect of evaporation on the fluid flow can be significantly decreased. Finally, a device with three channels converging to a single reaction zone was fabricated and sequential delivery of fluid to the reaction zone was demonstrated. The devices made by the methods described in this paper offer significant benefits for in-field diagnostic testing where multistep protocols are required.

A device fabricated by this method is estimated to cost between 10 and 50 ¢ and the plasma processes used to fabricate these devices are compatible with roll-to-roll technology, making it suitable for large scale production. This alternative fabrication approach contributes significantly toward commercialization of low-cost μ-PADs that meet the WHO ASSURED criteria.

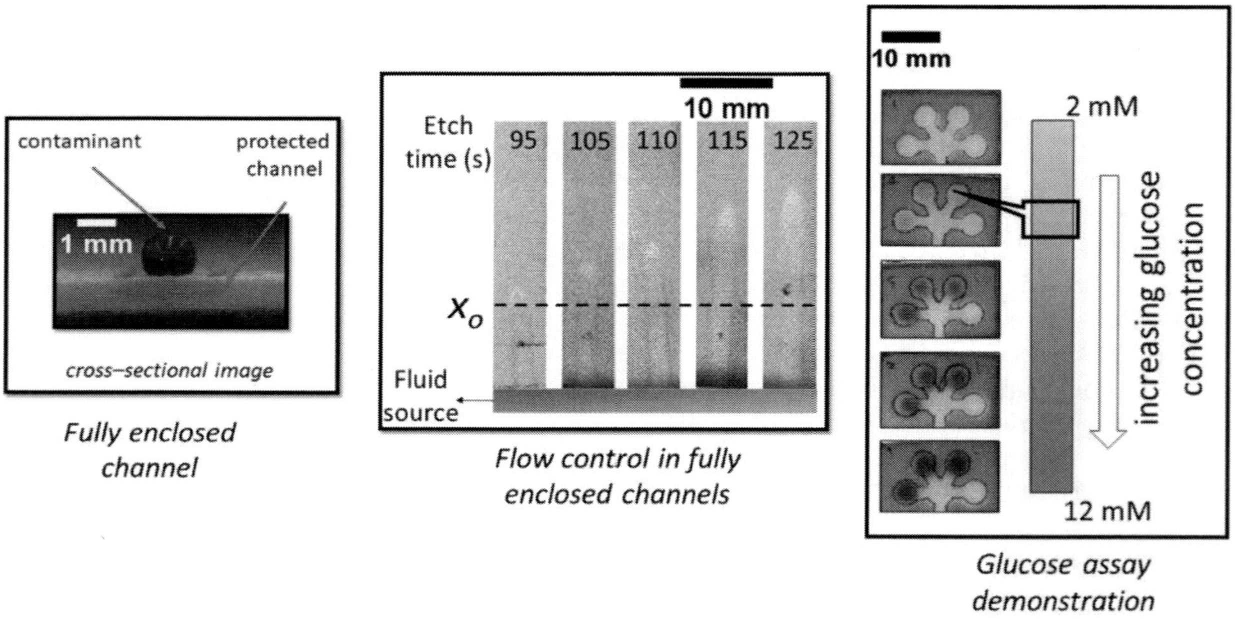

Copyright © American Institute of Chemical Engineers. All rights reserved.

611797 *Design of Gas-Liquid Transfer Apparatus for Concentration and Detection of Water Soluble Tuberculosis Associated Volatile Organic Compounds.*

Abstract:

Many human diseases have corresponding volatile organic compound (VOC) profiles in breath and breath analysis is increasingly the focus of new diagnostic techniques. This diagnostic approach is non-invasive and can serve as a safer, less time intensive diagnosis tool than blood tests. A number of diseases have been found to have breath profiles unique to the disease; such as influenza, diabetes, pneumonia, and tuberculosis (TB).

One of the principle issues with using breath profiles for diagnosis is that breath samples contain a complex array of more volatile and reactive compounds that often confound the discovery of relevant diagnostic biomarkers. Additionally, there are significant design challenges that make maintaining consistent sampling procedures and sample storage difficult. This consistency is essential for results comparison across samples.

With these issues in mind, a method of gas to liquid sample transfer is needed that provides consistent sampling results, improves sample stability during storage, enables the use of well-established electrochemistry techniques and produces a simpler, clearer breath profile with many of the more volatile, confounding VOCs removed. Preliminary data showing the effectiveness of this method can be seen in figure 1 shown below. While confounding VOCs were removed, the signature biomarker needed for disease diagnosis was still present; enabling clear identification as seen in figure 2. Details showing the apparatus for processing breath and results will be presented.

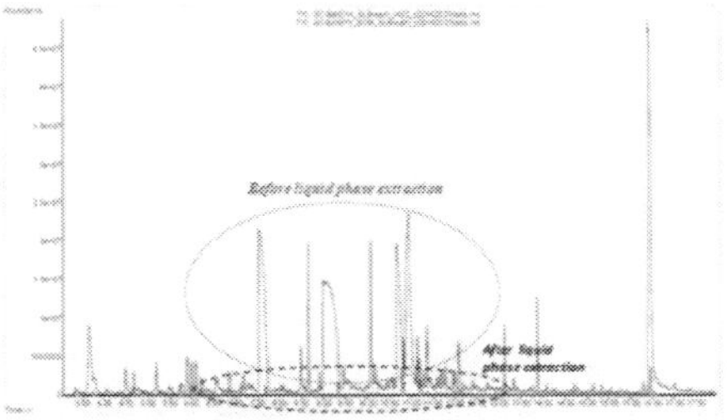

Figure 1: Superimposed total ion chromatograms of breath sample before and after liquid phase extraction.

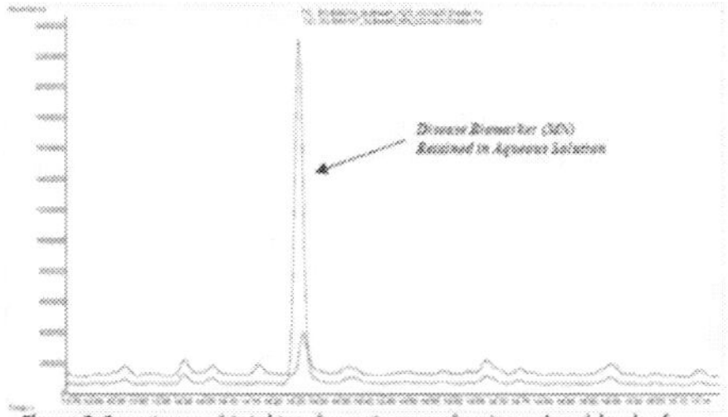

Figure 2: Superimposed total ion chromatogram showing reduced levels of compounds. Signature disease biomarker is maintained at approximate elution time of 14.26 minutes

Copyright © American Institute of Chemical Engineers. All rights reserved.

597539 Designer Mucus: Poolable and Tunable Airway Mucus for Biochemical, Biophysical, and Transport Studies

Abstract:

Mucus is the first barrier of defense in the human airway system. It is a biopolymeric gel composed of the gel-forming mucins MUC5B and MUC5AC as well as thousands of other biomacromolecules and tightly regulated ionic and water compositions. As a viscoelastic hydrogel its rheology is finely tuned to facilitate clearance from the airways via the mucociliary escalator. Biochemically, mucus is highly adherent to inhaled particles and pathogens, which are subsequently bound to and cleared with the mucus from the lungs. Pathology – particularly hyperconcentration – of mucus occurs in diseases like chronic bronchitis, asthma, and cystic fibrosis, affecting millions of people worldwide. Viral, bacterial, and particulate evasion of the airway mucus barrier affect millions more.

In our work we exploit the fine tuning of the physiological composition of airway of mucus to generate tunable pools of "stock" mucus from heterogeneous mucus samples obtained from living human airways following intubation as part of routine surgery. We demonstrate that compared to "isotonic" samples (i.e. samples obtained with normal ionic composition) that we have previously characterized, samples that were either dehydrated (i.e. hypertonic, due to air exposure) or diluted (i.e. hypotonic, possibly due to salivary contamination) can be either adjusted or combined into pools having physiological composition. These pooled stocks demonstrate bulk and microrheological properties similar to isotonic ETT mucus and mucus from other, non-airway, sources. The composition of the mucus stock can be altered to increased solids concentration and additional free DNA and other components found in mucus during acute or chronic airway disease to recapitulate the biochemistry and biophysics of sputum (e.g. from patients with cystic fibrosis). Furthermore, these stocks can be used to study particle transport through native, physiologically healthy mucus on the size scale of hazardous particles (e.g. asbestos), bacteria, and virions (approximately 10, 1, and 0.1 μm, respectively). Our results demonstrate the utility of native mucus "engineered" to reflect the physiology of the airway mucus barrier in health and disease.

Copyright © American Institute of Chemical Engineers. All rights reserved.

600074 Hydrodynamic-Induced Stress Distribution Experienced By VWF Protein Models

Abstract:

The von Willebrand factor (vWF) is a blood protein that is related to hemostasis. Changes to the protein structure can cause disease and difficulty for blood to clot, a disease known as the von Willebrand syndrome [1-3]. Hydrodynamic conditions can lead to flow-induced stress on the protein that can change its structure, when the flow is not physiological. Such non-physiological flow conditions occur when blood flows through vascular implants or other devices, like blood pumps. In this work, we use direct numerical simulations (DNS) of turbulent blood flow and transitional blood flow to calculate the distribution of stresses on particles that serve as models of individual vWF proteins. A case of fully developed flow and a case of Poiseuille-Couette flow are investigated. The protein motion in the flow field is calculated by Lagrangian Scalar Tracking (LST) of thousands of protein molecules that behave as passive particles in the flow field [4, 5]. These molecules have diffusivity that corresponds to the Stokes_Einstein diffusivity for vWF molecules. The vWF particles are released at different locations in the flow field, including the viscous wall subregion, the buffer region, the log layer and the outer region of the flow. The LST allows the calculation of the shear stresses acting on the vWF along its trajectory. The distribution of the stresses as a function of time and as a function of the point of release is calculated in both flow cases. The fully developed turbulent flow can occur when blood moves through a blood pump, at Reynolds numbers comparable to those suggested in the FDA Critical Path Initiative [6], where the computational fluid dynamics community was asked to model blood flow and hemolysis. The Poiseuille-Couette flow occurs when blood moves through a Ventricular Assist Device (VAD) heart implant [7]. It is found that both the flow field and the location of vWF release are important for the shear stress distribution. We will discuss the methodology for calculating the stresses in the Lagrangian framework and the effects that the history of the stresses can have on vWF configuration within the flow field. Since the stress on the protein is a function of time, the history of stresses experienced by a vWF molecule is important, instead of only the average stress or the wall shear stress that is used conventionally.

Literature Cited:

[1] Schneider, S.W, Nuschele, S., Wixforth, A., Gorzelanny, C., Alexander-Katz, A., et al. Shear-induced unfolding triggers adhesion of von Willebrand factor fibers. Proc. Natl. Acad. Sci. USA 104, 7899–7903 (2007)

[2] Lippok, S., Radtke, M., Obser, T., Kleemeier, L., et al. Shear-Induced Unfolding and Enzymatic Cleavage of Full-Length VWF Multimers, Biophysical J. 110, 564-554 (2016)

[3] Bekard, I.B., Asimakis, P., Bertolini, J., & Dunstan, D.E. Review of the effects of shear flow in protein structure and function. Biopolymers 95(11) 733-745 (2011)

[4] Nguyen, Q. and D.V. Papavassiliou. A statistical model to predict streamwise turbulent dispersion from the wall at small times. Physics of Fluids, 28(12), Art. 125103 (2016)

[5] Nguyen, Q., Feher, S., and D.V. Papavassiliou, Lagrangian Modeling of Turbulent Dispersion from Instantaneous Point Sources at the Center of a Turbulent Flow Channel, Fluids, 2(3), Art. 46 (2017)

[6] Malinauskas R.A., Saha A., Sheldon M.I. Working with the Food and Drug Administration's Center for Devices to advance regulatory science and medical device innovation. *Artif Organs*; 39:293-9, 2015.

[7] Coghill, P.A., Suren, K., Zheila J.A-N, Long, J.W. & Snyder T.A. Benchtop von Willebrand Factor Testing Comparison of Commercially Available Ventricular Assist Devices and Evaluation of Variables for a Standardized Test Method. ASAIO Journal,65(5):481-488, 2019.

Copyright © American Institute of Chemical Engineers. All rights reserved.

601116 Personalized Medicine for in Vitro Fertilization (IVF) Procedure with Ganirelix Protocol

Abstract:

In-vitro fertilization (IVF) is the most common technique in assisted reproductive technology. It has four basic stages: superovulation, egg retrieval, fertilization, and embryo transfer. Superovulation is a drug-induced method to enable multiple ovulation per menstrual cycle and key component towards a successful IVF cycle. Although there are the general guidelines for dosage, the dose is not optimized for each patient.

A mathematical procedure is developed based on analogy between crystallization and superovulation stage in IVF treatment. A moment model which describes the dynamics of follicle size distribution is derived. This model can provide a customized model of this stage regarding the size distribution of eggs (follicles/ oocytes) obtained per cycle as a function of the chemical interactions of the drugs used and the conditions imposed on the patient during the cycle, which provide a basis for predicting the possible outcome. This model and optimal control procedure were previously developed for one of the agonist dosage protocols used in clinical practice. Ganirelix is used in conjunction with FSH to prevent premature ovulation and results in a shorter treatment protocol which produces high quality eggs and is known to reduce occurrence of Ovarian Hyperstimulation Syndrome. This paper uses the same principles to model an antagonist dosage protocol using Ganirelix (GnRH) and the optimal control procedure for improving outcomes of IVF treatment. However, in case of Ganirelix, two more reaction terms are added which corresponds to change in growth rate due to introduction of Ganirelix in mid cycle.

Customized patient-specific model parameters are obtained by using initial two-day data for each patient. Subsequently, this model is used to predict the FSD for the remaining days of the cycle. This procedure was conducted for 13 patients. The results of the customized models are found to be closely matching with the observed FSD. Rise in estrogen levels indicate the successful progress of cycle and estrogen level determine quality of eggs. We found that the estrogen level is a linear function of third moment of follicle size distribution at any time. A good fit was observed between the customized patient model and the experimental estrogen data, thus indicating a successful model. Optimal control is then used to find personalized optimal dosage for each patient for each day of cycle.

Copyright © American Institute of Chemical Engineers. All rights reserved.

609196 Computation-Driven Analysis of Model Polyextremotolerant Fungus *Exophiala Dermatitidis*: Defensive Pigment Metabolic Costs and Human Applications

Abstract:

The polyextremotolerant black yeast *Exophiala dermatitidis* is a tractable model system for investigation of adaptations that support growth under a variety extreme conditions such as pH, light, radiation, anti-fungal compounds, temperature, and more. Foremost amongst these adaptations is melanogenesis and carotenogensis, which produce the two primary defensive pigment classes in *E. dermatitidis*: melanins and carotenoids. A particularly important question is their metabolic production cost, which can give insight into the costs and trade-offs associated with and limits to polyextremotolerance. However, investigation of this issue has been hindered by a relatively poor systems-level understanding of the *E. dermatitidis* genome and metabolism. To address this challenge, a genome-scale model (*i*Ede2091) was developed using a variety of tools to overcome current knowledge gaps such as BLASTp and OptFill. By applying the concept of shadow price to the production of pigments in *i*Ede2091, carotenoids were found to be more expensive to produce than melanins. Given that all functions performed by carotenoids are also known to be performed by melanins and that melanins are the first line of defense, this suggests that carotenoids have an underexplored yet important role, possibly in photo-protection. Furthermore, given that defensive pigments with overlapping functions generally have different shadow prices under different growth conditions, this might indicate that *E. dermatitidis* may minimize the cost of polyextremotolerance through selective production of the least expensive pigment. In addition to shadow price analyses, *i*Ede2091 revealed that *E. dermatitidis* synthesizes the same melanins as humans and the active sites of the key tyrosinase enzyme are highly conserved this model may enable a broader understanding of melanin production across kingdoms.

Publication: W. L. Schroeder, S. D. Harris, and R. Saha. "Computation-driven analysis of model polyextremotolerant fungus *Exophiala dermatitidis*: defensive pigment metabolic costs and human applications". *iScience*, vol. 23, no. 4, Apr. 24, 2020. Available: https://www.cell.com/iscience/fulltext/S2589-0042(20)30164-4 (doi: https://doi.org/10.1016/j.isci/2020.100980)

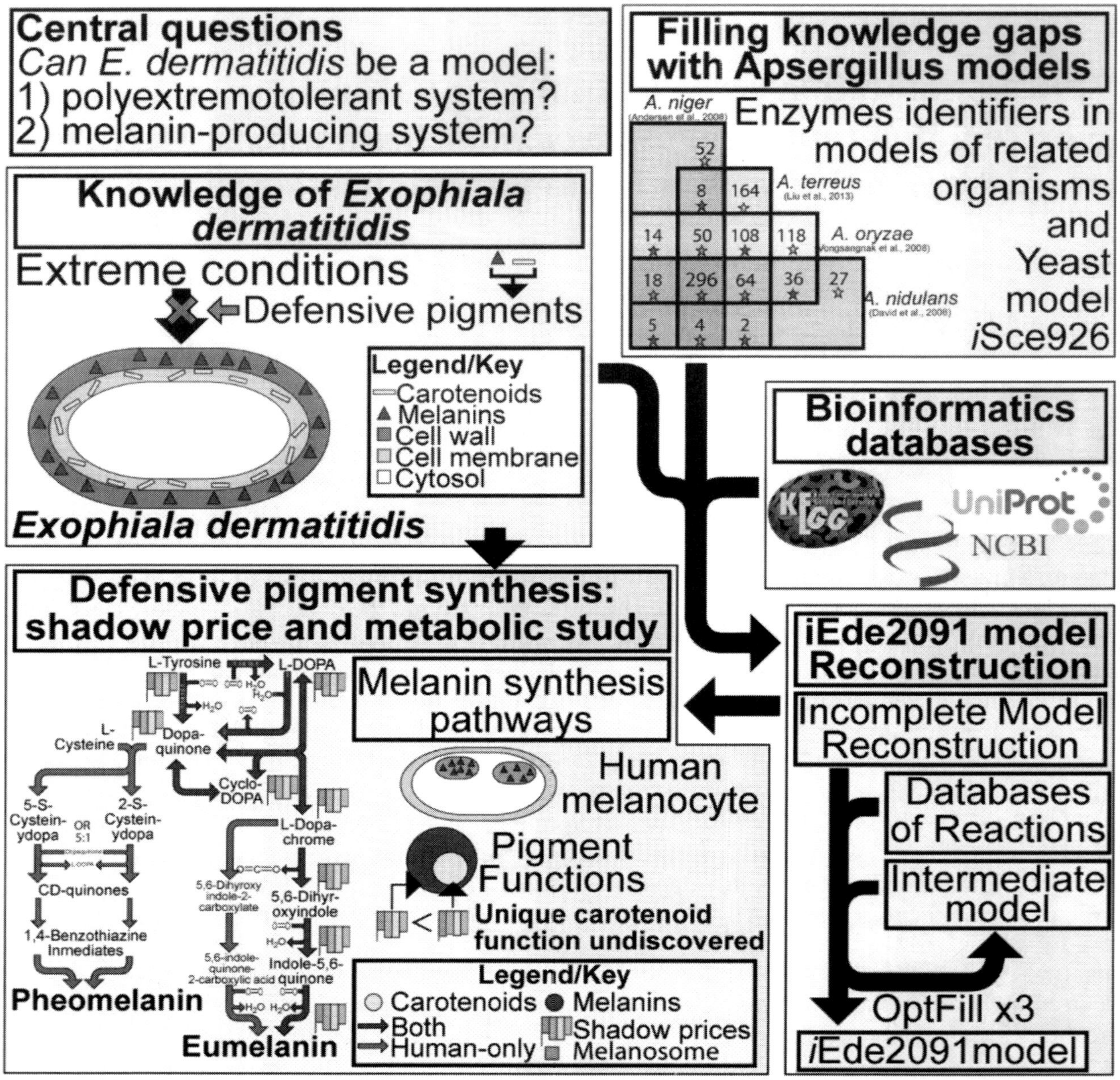

Copyright © American Institute of Chemical Engineers. All rights reserved.

608064 Biohybrid Polymersome-Mediated Enzyme Replacement Therapy for GM1 Gangliosidosis

Abstract:

Background: Lysosomal storage disorders (LSDs) are a group of 50 rare metabolic disorders caused by deficiency or absence of the lysosomal enzymes and accumulation of their substrate molecules in lysosomes. Enzyme replacement therapy (ERT) is one of the currently available treatments of LSDs. It involves intravenous delivery of the missing enzyme from an external source to compensate for the deficiency. GM1 gangliosidosis is an autosomal recessive LSD caused by the mutations in the GLB1 gene, which encodes the production of the lysosomal enzyme β-galactosidase (β-gal). Subsequently, substrate molecules of β-gal, primarily GM1 ganglioside, and others like keratan sulfate and a few oligosaccharides with terminal β -galactosyl groups build-up in the lysosomes[1]. GM1 gangliosidosis has a prevalence of one in 100,000 to 200,000[2]. Due to the CNS involvement in GM1 gangliosidosis, the blood-brain barrier (BBB) presents a challenge for treatment using ERT. The presence of BBB necessitates the use of a drug delivery system that can encapsulate and transport the deficient enzyme across the BBB through receptor-mediated transcytosis (RMT). Previous work in our lab[3] demonstrates the effectiveness of a pH-responsive polymer-based system to encapsulate and deliver β-gal to GM1 gangliosidosis-affected cells. However, the payload release is limited to ~35%.

Motivation: A lysosomal enzyme β-hexosaminidase A (HexA) is previously reported to be upregulated in GM1 gangliosidosis-affected felines, with HexA reaching normal levels after treatment with a viral vector[4]. To utilize this phenomenon, hyaluronic acid-*b*-poly(lactic acid) (HA-PLA) nanoparticles are used to encapsulate and deliver β-gal. HexA degrades hyaluronic acid[5], and PLA undergoes acidic hydrolysis, thereby rendering the nanoparticles dual-responsive. Due to the presence of hydrophilic HA and hydrophobic PLA, the amphiphilic HA-PLA copolymer readily self-assembles into nanoparticles in the aqueous phase. The hydrophilic fraction (*f*) of a polymer is the ratio of the mass of hydrophilic block to the total mass. It dictates the structure of the self-assembled nanoparticles. The enzymatic nature of the cargo requires the use of vesicular nanoparticles called polymersomes. Functionalizing the surface of these nanoparticles with apolipoprotein E (ApoE), a ligand that binds to the low-density lipoprotein receptors expressed on the BBB, can facilitate the RMT across BBB[6,7].

Methods: The HA-PLA polymer is synthesized through a two-step polymer conjugation process[8]. Briefly, hyaluronic acid (mol. wt. 5000 Da) is terminally functionalized with 1,4-diaminobutane to introduce an amine group at the reducing end. The aminated HA is then reacted with N-Hydroxysuccinimide (NHS) activated PLA in the presence of N,N-diisopropylethylamine. The reaction mixture is dialyzed against DI water to remove solvents and unreacted hyaluronic acid, followed by lyophilization to obtain the polymer HA-PLA. The polymer is subsequently used to synthesize nanoparticles using the solvent-injection method. The particle size and structure are analyzed using the dynamic light scattering (DLS) and transmission electron microscopy (TEM), respectively. The enzyme-based degradation of nanoparticles is determined using both DLS and TEM.

Results: The conjugated polymer is analyzed using Attenuated total reflectance-Fourier-transform infrared (ATR-FTIR) spectroscopy. The resultant spectrograph analysis demonstrated peaks corresponding to hyaluronic acid and poly(lactic acid) polymer blocks, confirming the conjugation.

To prepare nanoparticles, HA-PLA is dissolved in dimethyl sulfoxide and injected into aqueous phase at controlled rate. DLS analysis of nanoparticles after filtration resulted in a peak diameter of 199.4 ± 42.6 nm. The TEM imaging of the nanoparticles revealed a vesicular structure. To determine the enzyme-based degradation of the polymersomes, they were incubated separately with hyaluronidase (HYAL) (cognate enzyme) and β-gal (non-cognate enzyme). HexA and HYAL have similar functionality with regards to the degradation of hyaluronic acid[5]. After incubating with the enzymes for 24 hours, particles incubated with HYAL presented a peak diameter (in particle size distribution of DLS analysis) at a smaller size indicating degradation. The peak diameter of the particles incubated with β-gal remained in the same range as before incubation indicating little to no degradation. For release study, polymersomes loaded with the dye Fluorescein Isothiocyanate-Dextran (FITC-D) as model encapsulant were incubated with HYAL and β-gal, and the fluorescence signal is monitored for hours. A higher release of FITC-D upon incubation with HYAL was observed compared to the release in β-gal incubation.

Conclusions: These results corroborate the conjugation of the hyaluronic acid and poly(lactic acid) blocks. The ability of the amphiphilic HA-PLA polymer to self-assemble into vesicular nanoparticles has been proven with the help of TEM. DLS analysis and TEM imaging show that the resultant nanoparticles can undergo enzymatic degradation in the presence of hyaluronidase. Release study showed the selective degradation of the nanoparticles in the presence of cognate enzyme (HYAL).

References:

1 Sandhoff, K. *et al. J. Neurosci.* 33, 10195–10208 (2013)

2 Kelly, J. M. *et al. Progress in Neurobiology* 152, 166–180 (2017)

3 Kelly, J. M. *et al. Nanomedicine* 12, 2591–2606 (2017)

4 McCurdy, V. J. *et al. Sci. Transl. Med.* 6, 231–279 (2014)

5 Gushulak, L. *et al. J. Biol. Chem.* 287, 16689–97 (2012)

6 Dehouck, B. *et al. J. Cell Biol.* 138, 877 LP – 889 (1997)

7 Wang, D. *et al. Proc. Natl. Acad. Sci. U. S. A.* 110, 2999–3004 (2013)

8 Deng, C. *et al. J. Mater. Chem. B* 6, 3163–3180 (2018)

Copyright © American Institute of Chemical Engineers. All rights reserved.

599586 Metaheuristic Optimization of Graphical Testing Procedures for Clinical Trials

Abstract:

When evaluating the safety and efficacy of compounds during a clinical trial, there is always the risk that a dataset will produce a statistically significant indication despite being the result of random noise, termed a "type-1 error." Furthermore, when multiple hypotheses are tested in the same study, the likelihood of producing any spurious indications grows unless appropriate steps are taken to control the family-wise type-1 error rate. A commonly used approach is to commit to a graphical hypothesis testing procedure before analyzing the trial results. This involves allocating a "budget" of statistical significance to each clinical effect and committing to procedures for "sharing" significance in a way that strongly controls the family-wise type-1 error rate. Currently, there are a variety of strategies for manually building a graphical hypothesis test, but there are no algorithmic approaches for rigorously optimizing the design of a graphical hypothesis test subject to the inherent uncertainty in clinical outcomes. In this work, we translate the problem of designing graphical hypothesis tests to suitably defined mathematical optimization models. We then demonstrate approaches for systematically optimizing the resulting models. Due to the presence of parametric uncertainty and the highly non-convex nature of the graphical hypothesis testing procedures, we employ metaheuristic optimization algorithms and provide a thorough characterization of their algorithmic performance. This work streamlines the process of identifying clinical trial designs that are both powerful and averse to risk.

Copyright © American Institute of Chemical Engineers. All rights reserved.

601145 Multiple Particle Tracking Detects Changes in Brain Extracellular Matrix Structure

Abstract:

Motivation: Brain extracellular matrix (ECM) structure mediates many aspects of neuronal function. For example, the formation of highly condensed perineuronal nets (PNNs), which envelop the soma of certain populations of neurons in the brain, helps regulate plasticity and protects neurons from harmful processes like excitotoxicity and oxidative stress. Brain ECM is also highly dynamic, and the ability to assemble, disband, and reorganize is required for the development of proper neuronal circuitry and helps facilitate repair in response to injury. The structural integrity of PNNs is thought to be impacted by many neurological diseases, including epilepsy, schizophrenia, and stroke. However, probing real-time changes in ECM structure, particularly changes that occur at the cellular level in living tissue, remains an ongoing challenge. This prevents a complete understanding of the role disease-induced changes in ECM structure play in impairing neuronal function. Herein, we introduce a novel approach we have taken to address this current knowledge gap and present our initial findings to demonstrate its utility.

Approach: Our approach uses multiple particle tracking (MPT), a technique that leverages fluorescent microscopy to capture the motion of nanoparticles in real-time. MPT is unique in that the microscopic behavior of hundreds to thousands of individual particles can be tracked simultaneously, while retaining single particle resolution. The motions exhibited by particles provide information about the environment in which they reside, and the ability to track the movement of individual particles provides high spatial resolution. By using sub 60 nm tracer particles with polyethylene glycol (PEG) surface coating, we are able to capture diffusion within the narrow channels of the brain extracellular space (ECS) and evade adhesive interactions with cellular components. The motion of these particles is thus influenced by local fluid properties of the ECS and structural properties of the local ECM. To determine whether MPT was sensitive enough to detect changes in local ECM structure, we applied it to two separate scenarios *ex vivo*. First, we enzymatically-induced changes in ECM structure in rat brain slices by treating them with one of two ECM degrading enzymes. We then tested whether MPT could detect a rearrangement of brain ECM that occurs naturally during neurodevelopment. For this, MPT experiments were performed in brain slices taken from rats spanning a range of ages within the critical period of neurodevelopment.

Materials and Methods: Acute whole hemisphere rat brain slices were prepared by taking 300 μm-thick coronal sections from freshly euthanized Sprague-Dawley rats. Brain slices were cultured on membrane inserts at physiologically relevant conditions. *Ex vivo* experiments were split into two groups: (1) a group that applied exogenous enzymes to induce changes in brain ECM structure, and (2) a group that relied on a rearrangement of brain ECM that occurs naturally during development. For group (1), litter matched brain slices were collected from rats on postnatal days 35, 36, 37, and 38 (P35-P38). Within 24 h of euthanasia, slices were treated with either chondroitinase ABC (ChABC, 0.4 U/mL), hyaluronidase (HYase, 35 U/mL), or culture media. Treatments lasted for 2 h, at which point MPT was performed. 30 min prior to video acquisition, brain slices were injected with 51 nm PEG-coated fluorescent polystyrene nanoparticles (PS-PEG). Videos of nanoparticle displacements were taken from the cortex of each slice at 33 frames-per-second and 100x magnification for 651 frames using a cMOS camera mounted on a confocal microscope. Nanoparticle trajectories, trajectory mean squared displacements (MSDs), and effective diffusion coefficients (D_{eff}) were calculated using a Python package developed within our group. For group (2), slices were collected from P14, P21,

P28, and P35 rats. Within 24 h of slice collection, the same 51 nm PS-PEG nanoparticles were injected into brain slices and MPT was carried out in the cortex. Videos were collected and results generated using the same methodology. In both instances, a change in the presence of PNNs was used as an indicator of changes in brain ECM structure. To detect changes in the presence of PNNs, brain slices were stained with fluorescein-labeled wisteria floribunda agglutinin (WFA) lectin at 10 μg/mL for 12 h at 4°C and imaged using a confocal microscope.

Results: Both treatments, ChABC and HYase, led to a complete loss of PNNs in P35 rat brain slices within 2 h (Figure 1). PNNs present in slices treated with culture media were not affected. Nanoparticles explored a greater area, moved faster, and had increased diffusivities when diffusing in enzyme-treated brain tissue. Despite having the same number of total trajectories (1478 and 1732 for non-treated and treated, respectively), nanoparticles surveyed a greater fraction of the ECS when diffusing in the slice treated with an ECM-degrading enzyme. In enzyme-treated slices, geometrically averaged mean-squared displacements (<MSD>) of nanoparticle trajectories were greater in magnitude for all lag times between 0 and 2 s. Lastly, nanoparticles diffusing in treated slices had significantly larger D_{eff} values than those diffusing in non-treated slices. We then transitioned to the age dependent study. PNNs stained using WFA did not appear qualitatively until 21 days after birth (P21), and preferentially formed around parvalbumin-expressing (PVA^+) interneurons. The areal density of PNNs in the cortex increased throughout development, with significant differences existing between P14 and P28, P14 and P35, and P21 and P35 ($p < 0.05$) groups. Tracking revealed an inverse relationship between nanoparticle diffusive ability and brain age. <MSD> profiles decrease in magnitude as pup age increases from 14 days to 21, 28 and 35 days after birth. Distributions of D_{eff} values at a 0.33 s lag time shift to lower values as the brain develops and correspondingly PNN density increases.

Conclusions: Brain ECM plays critical roles throughout neurodevelopment, helps preserve long-term neuronal health, and facilitates repair in response to injury. Altered ECM structure is thought to be involved in the pathophysiology of many neurological diseases, and the implications of altered PNN integrity on neuronal plasticity and activity has garnered significant attention in recent decades. Until now, probing real-time changes in ECM structure, especially changes occurring at the cellular level in living tissue, has been challenging. Our findings suggest that MPT is capable of this. In addition to leveraging the brain's natural tendency to restructure during the critical period of development, we also applied MPT to *ex vivo* hemispheric brain slices undergoing an enzymatically induced breakdown of ECM. In both instances, significant differences in nanoparticle diffusive ability existed between experimental groups. The further application of MPT in studying ECM structure could more explicitly define mechanisms involved in neurological disease progression and open new avenues of therapeutic intervention. Additionally, MPT can enhance our baseline understanding of the structure-function relationships of the brain under normal physiological conditions and has the potential to become used as one marker of neurological disease severity.

Figure 1. MPT detects enzymatically-induced breakdown of PNNs in rat brain slices *ex vivo*. (A) Schematic representation of PNN breakdown following treatment with HYase or ChABC. (B) Representative 20x magnification images taken from the cortex of P35 rat brain slices either treated with an ECM-degrading enzyme or not treated. Rows represent treatment group. Columns represent treatment time. (C) Representative trajectory maps generated from MPT experiments carried out in non-treated (blue) and HYase-treated (grey) P35 brain slices *ex vivo*. (D) Geometrically ensemble averaged <MSD> versus lag time quantified from trajectories of nanoparticles diffusing in non-treated (blue), ChABC-treated (gold), and HYase-treated (grey) P35 brain slices. Faint lines represent individual videos (60 total per group). Bold line represents the mean of 60 videos. (E) Scatter plots of D values generated from all videos (n = 5) in all slices (n = 3) in all brains (n = 4), separated by eff

treatment group. Bars show median value (95% CI). * denotes significant differences (Kruskal-Wallis test) between groups, after adjusting for multiple comparisons. ($p < 0.05$)

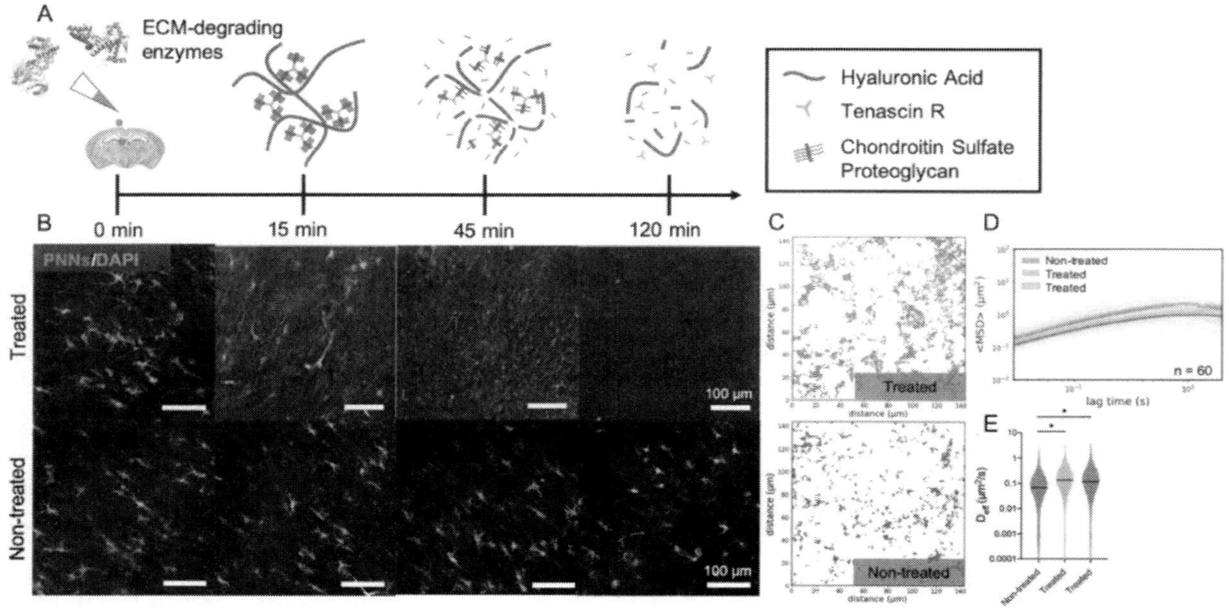

Copyright © American Institute of Chemical Engineers. All rights reserved.

601987 Producing Predictive Chemical Process Models from a State of Incomplete Mechanistic Knowledge and Small Data Sets

Abstract:

ABSTRACT: In this work, we explore the possibility of developing a quantitative and predictive model for chemical processes, without having to first gain a complete understanding of the underlying mechanisms. We find that this can be done efficiently combining: (1) expectations for the functional expression of the process—extracted from a general, *qualitative* understanding on how the chemical process ought to behave—and, (2) small data sets providing examples of the process outputs resulting from varying process inputs. The strategy discussed is labeled knowledge-constrained machine learning. The main example we take is on reaction kinetic models—and, here, the expectations are written in terms of constraints on the model predictions over time, and the data is a series of time-point measurements for the reaction outcome under different conditions. But, the intent of the presentation is to discuss a general methodology that may be used to develop predictive models across the range of unit operations found in the processes used to produce synthetic drug substance API's.

Copyright © American Institute of Chemical Engineers. All rights reserved.

594384 Identifying the Processing Space of Continuous Granulation: A Case Study of Extended Release Tablets

Abstract:

While continuous manufacturing (CM) has been employed for directly compressible tablets, the continuous wet granulation has also been used as an alternative in continuous tableting lines. To date, no matrix-based extended release tablets manufactured by a CM technology has been approved. Compared to batch granulation, continuous twin screw granulation exerts higher shear to produce denser granules with broader size distribution within very short time (seconds). Changing the screw configuration, process, and formulation parameters can alter the physical properties of the granules, e.g., shape, flow, compressibility, density, size distribution, etc. In twin screw granulation, complex interactions among these parameters manage the granule properties. This study aimed at identifying the critical material attributes and process parameters and their interactions to be controlled for switching a batch process to a CM granulation process.

Using batch high shear granulation, we evaluated the formulation and process parameters that can produce metoprolol succinate extended release tablets. These formulation and process parameters were successfully transferred to continuous twin screw granulation to produce the same dissolution properties of the finished product. The effects of the screw configuration parameters on the characteristics of the granules were further investigated. The interaction between the processing, e.g., liquid to solid ratio, screw speed, barrel fill volume, etc., and the significant screw parameters was studies to unveil the processing space and to identify the failure modes.

Similar granule quality attributes were obtained for most of the formulations; although, the conditioned bulk density and compressibility of the granules determined the dissolution properties of the tablets. The number of kneading and sizing elements had the most significant impact on granule characteristics. The effects of the staggering angle and distance between the kneading zone of the screws were negligible.

This presentation will then highlight the processing space of screw granulation to produce extended release tablets. Some of the failure modes will be discussed along with the feasibility of implementing PAT tools to monitor the granulation process and granules properties.

DISCLAIMER

This presentation reflects the views of the presenter and should not be construed to represent FDA's views or policies.

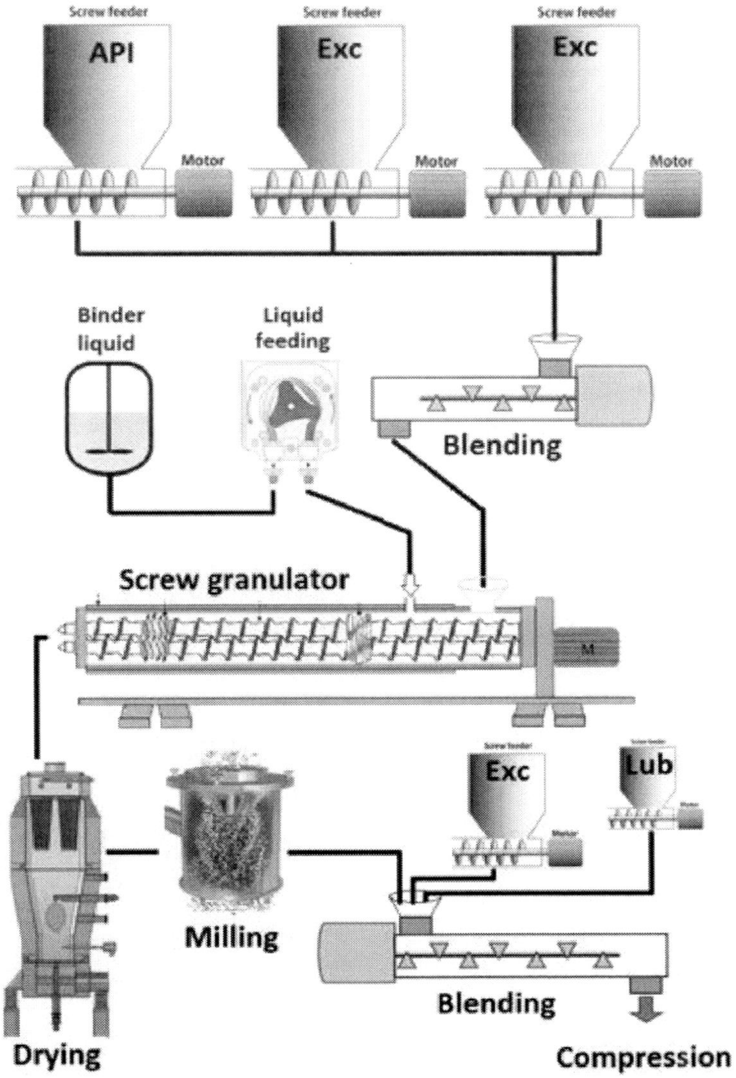

Copyright © American Institute of Chemical Engineers. All rights reserved.

608665 Dynamic Optimization Strategies Towards Addressing the Challenges in Protein Therapeutics

Abstract:

Protein therapeutics is an important field of the pharmaceutical industry. Current challenges in this field include meeting the increasing market demands, handling the entrance of biosimilar products and complying with quality by design (QbD) principles [1-3]. To tackle these challenges effectively, efficient use of existing tools, methods and solution approaches together with the development of novel ones is necessitated. This work provides an overview of our recent efforts to address some of the aforementioned challenges using model-based dynamic optimization strategies.

Model-based approaches are powerful tools towards facilitating process development and advancing process performance. They offer a promising alternative to costly and time-consuming experimentation [1], and can have a profound effect on process sustainability and profitability. Recent advances in cell culture modeling (with cell cultures representing a typical production platform for many of the products of this class) have enabled significant insight to the system's dynamics and the underlying production mechanisms, and have paved the way towards developing dynamic optimization strategies for process intensification (cf., [4-6]).

Existing numerical methods and optimization software for solving the resulting dynamic optimization problems rely almost exclusively on local techniques. Due to the noncovexity of the described complex biological systems, the vast majority of these problems exhibits the presence of multiple local minima. Nevertheless, deriving suboptimal local solutions can have direct quality and cost impacts. Despite the significant progress in deterministic global optimization [7], it is not yet tractable for the considered systems. Hence, alternative methods to overcome the hazards of suboptimal local optimality are required. Stochastic and heuristic approaches can be advantageous to this end. In our previous work [8], we have shown how incorporating biological process knowledge into the optimization problem formulation can facilitate the derivation of superior local solutions. In a similar direction, we illustrate how proper decomposition of the optimization problem can outweigh certain advantages [9].

Beyond increasing product yield, monitoring and controlling product quality is essential. Regulatory authorities including FDA are increasingly recommending consideration of QbD to develop processes that consistently meet quality specifications [3, 10]. This requires further attempts in modeling, as well as reconsiderations in the formulation of the dynamic optimization problem to effectively account for direct quality attributes. Yet, current methods in this direction are still in an early stage. The recent work in [11] demonstrates a successful example of model-based optimization to maximize galactosylation (a glycosylation related metric, with glycosylation presenting a critical quality attribute for many therapeutic proteins [12, 13]) content for an antibody case study. In our recent study [14], we have combined model simulations and design of experiments to investigate the effects of nutrients and nucleotide sugars feeding in protein glycosylation. Building on this basis, we have developed an optimization framework to maximize process performance while meeting quality specifications, and thus provided a successful example for utilizing dynamic optimization to apply the QbD paradigm in protein therapeutics [15].

Acknowledgements:
This work has received funding from the European Union's Horizon 2020 research and innovation programme under the Marie Skłodowska-Curie grant agreement no.675251.

References:
[1] Farzan, P., Mistry, B., & Ierapetritou, M. G. (2017). Review of the important challenges and opportunities related to modeling of mammalian cell bioreactors. AIChE Journal, 63(2), 398-408.
[2] Shukla, A. A., Wolfe, L. S., Mostafa, S. S., & Norman, C. (2017). Evolving trends in mAb production processes. Bioengineering & Translational Medicine, 2(1), 58-69.
[3] Collins, P. C. (2018). Chemical engineering and the culmination of quality by design in pharmaceuticals. AIChE Journal, 64(5), 1502-1510.
[4] Dhir, S., Morrow Jr, K. J., Rhinehart, R. R., & Wiesner, T. (2000). Dynamic optimization of hybridoma growth in a fed-batch bioreactor. Biotechnology and Bioengineering, 67(2), 197-205.
[5] Kiparissides, A., Pistikopoulos, E. N., & Mantalaris, A. (2015). On the model-based optimization of secreting mammalian cell (GS-NS0) cultures. Biotechnology and Bioengineering, 112(3), 536-548.
[6] Kappatou, C. D., Mhamdi, A., Campano, A. Q., Mantalaris, A., & Mitsos, A. (2018). Model-based dynamic optimization of monoclonal antibodies production in semibatch operation—Use of reformulation techniques. Industrial & Engineering Chemistry Research, 57(30), 9915-9924.
[7] Chachuat, B., Singer, A. B., & Barton, P. I. (2006). Global methods for dynamic optimization and mixed-integer dynamic optimization. Industrial & Engineering Chemistry Research, 45(25):8373–8392.
[8] Kappatou, C. D., Mhamdi, A., Campano, A. Q., Mantalaris, A., & Mitsos, A. (2017). Dynamic optimization of the production of monoclonal antibodies in semi-batch operation. In Computer Aided Chemical Engineering (Vol. 40, pp. 2161-2166). Elsevier.
[9] Kappatou, C. D., Altunok, O., Mhamdi, A., Mantalaris, A., & Mitsos, A. (2019). Sequential and Simultaneous Optimization Strategies for Increased Production of Monoclonal Antibodies. In Computer Aided Chemical Engineering (Vol. 46, pp. 1021-1026). Elsevier.
[10] Pais, D. A., Carrondo, M. J., Alves, P. M., & Teixeira, A. P. (2014). Towards real-time monitoring of therapeutic protein quality in mammalian cell processes. Current Opinion in Biotechnology, 30, 161-167.
[11] Kotidis, P., Jedrzejewski, P., Sou, S. N., Sellick, C., Polizzi, K., del Val, I. J., & Kontoravdi, C. (2019). Model-based optimization of antibody galactosylation in CHO cell culture. Biotechnology and Bioengineering, 116(7), 1612-1626.
[12] Hossler, P., Khattak, S. F., & Li, Z. J. (2009). Optimal and consistent protein glycosylation in mammalian cell culture. Glycobiology, 19(9), 936-949.
[13] Blondeel, E. J., & Aucoin, M. G. (2018). Supplementing glycosylation: A review of applying nucleotide-sugar precursors to growth medium to affect therapeutic recombinant protein glycoform distributions. Biotechnology Advances, 36(5), 1505-1523.
[14] Ehsani, A., Kappatou, C. D., Mhamdi, A., Mitsos, A., Schuppert, A., & Niedenfuehr, S. (2019). Towards model-based optimization for quality by design in biotherapeutics production. In Computer Aided Chemical Engineering (Vol. 46, pp. 25-30). Elsevier.
[15] Kappatou, C. D., Ehsani, A., Niedenführ, S., Mhamdi, A., Schuppert, A., & Mitsos, A. Quality-targeting dynamic optimization of monoclonal antibody production. (Submitted)

Copyright © American Institute of Chemical Engineers. All rights reserved.

608965 Integrated Supply Chain Network Design and Inventory Management for Autologous Cell Therapy.

Abstract:

Personalized medicine is gaining a lot of attention in the pharmaceutical industry as it can treat diseases that earlier had no cure. In the past few years, we witnessed exponential growth in the development of cell and gene therapy products. Cancer Research Institute reported that 753 cancer cell therapies are currently under development all over the world, of which 375 are in clinical trials [1]. There has been an increase in the growth and demand for autologous therapy due to the recent approval from the FDA [2]. Therefore, autologous therapy is undergoing the transformation from laboratory to commercial scale.

The autologous therapy starts with a collection of cells from the patient, which are cryopreserved and sent to the manufacturing site. The cells are then formulated and enriched. The formulated cells are again cryopreserved and transferred back to the patient, where they are injected back. Unlike traditional pharmaceutical drugs that are produced in bulk, these drugs are patient-centric and have to be manufactured for each patient separately. In addition, the process also deals with living cells that go around different nodes in the supply chain. Therefore, the cells have to be cryopreserved in storage device and transported across the supply chain.

An integrated approach is considered for supply chain network design and inventory management of the raw materials. We consider a network with a set of hospitals where the patients arrive. The facilities to be set up are distribution centers and manufacturing centers. The manufacturing center receives the samples from the patient and engineers the cells to create the drug. The distribution centers have the inventory for the storage containers and are installed for the intermediate halt for the sample transportation between the hospital and manufacturing center. To account for the transit times for the material transfer between facilities, we define the parameter accounting for both transportation time and waiting time for different transportation modes as well as the processing times at each of the node. We also simultaneously optimize inventory policy. We use periodic-review policy for the raw materials inventory at the distribution center [3]. Under this policy, if the inventory is below the defined level, the order is placed to restore the inventory to level . Since the storage devices are reusable, the policy is modified to account for the returning devices after the delivery of the drug.

Given the demand distribution, the potential locations of distribution and manufacturing centers, costs and time for processing, inventory, and transportation, we determine the location of different centers, inventory levels at the distribution centers, response time, and cost of delivery for each sample, with objective function of minimizing the cost and response time. The problem is formulated as a single objective optimization where response time is multiplied by a factor and added to the cost function. A case study of China is considered, with 20 hospitals situated in different provinces and a total of 130 patients over a period of 1 month. Also, the location of 20 potential locations of the distribution centers and 20 manufacturing centers are given. We found out that the majority of the cost is incurred in setting up manufacturing and distribution centers. The cost accounting for the response time depends on the multiplicative factor we choose. In addition, we also explored the importance of batching the samples at the distribution centers and transporting to the manufacturing center subjected to the response time constraint. Decisions related to inventory of raw material are critical to have an efficient supply chain. Also, sensitivity analysis is conducted to study the effect of different parameters on the optimal network design and overall cost.

References:

1. Tang, J., et al., *The global landscape of cancer cell therapy.* Nat Rev Drug Discov, 2018. **17**(7): p. 465-466.
2. Bach, P.B., S.A. Giralt, and L.B. Saltz, *FDA approval of tisagenlecleucel: promise and complexities of a $475 000 cancer drug.* Jama, 2017. **318**(19): p. 1861-1862.
3. Brunaud, B., et al., *Inventory policies and safety stock optimization for supply chain planning.* AIChE Journal, 2019. **65**(1): p. 99-112.

Copyright © American Institute of Chemical Engineers. All rights reserved.

600490 Two-Stage Control Using Model-Based Reinforcement Learning and Predictive Control for Fed-Batch Bioreactor

Abstract:

As the demand for biomass and pharmaceutical materials is increasing in the market, various genetically modified cells are tested in a batch bioreactor in order to increase production rate [1]. At the same time, a robust and efficient control strategy is required to successfully produce the product with uncertainty of cellular behaviors [2]. However, limitation of the information on the system dynamics makes it challenging problems to optimize the operating conditions of a bioreactor and to implement an automatic controller. In general, the biosystem is less reproducible and shows higher nonlinearity compared to other engineering systems. In particular, most of the parameters in the first-principle model are lumped type and these values would be varied not only in a single batch operation with changes of the states but in a batch-to-batch fashion. Therefore, implementing an automatic control to track the optimal control trajectories obtained off-line is not a robust approach to produce fine products for a large number of batch reactors.

In this work, we propose a two-stage control strategy to optimally control a fed-batch type bioreactor that produces penicillin. The proposed strategy utilizes model-based Reinforcement Learning (RL) off-line to optimize the operating trajectories of state and input. In on-line, the Moving Horizon Estimation (MHE) is applied to estimate the parameter values, and an adaptive Model Predictive Control (MPC) is utilized to simultaneously modify and track the trajectories obtained in off-line. First, the first-principle model of the system dynamics is represented in Stochastic Differential Equations (SDEs) form which could theoretically handle the white noise in a continuous system. In addition, the variance of the process noise could shift a mean value of drifting terms of the system so that the changes in parameter values could be presented by adjusting the variance of the process noise [3]. These SDEs are numerically integrated and served as a virtual plant. The next step is obtaining the operating trajectories of the reactor off-line. The optimal control method could be classified in three classes, direct, indirect and RL (dynamic programming-based method). The direct method is the one mostly applied in practice since the problem formulation and the procedure of obtaining the solution is intuitive. On the other hand, RL has several advantages over the direct method. RL provides a closed-loop solution of the optimal control problem so that the control policy could be obtained instead of just a control trajectory [4]. This is a more robust approach since RL evaluates the control cost for the case where the states have deviated from the optimal trajectories. In addition, value based RL provides not only the optimal trajectories but also the value of state which is an expected total cost of the control policy starting from the state [4]. Although this value is just a by-product to find the optimal control policy, it is useful information when the trajectories are modified on-line.

The trajectories suggested by RL could serve as a baseline of optimal control. However, due to the change in parameter values and disturbances, tracking those trajectories might not be feasible which requires a modification of trajectories. For this purpose, MHE and adaptive MPC could be utilized to estimate the parameters and modify the trajectories on-line. However, even though changes in parameter values are successfully estimated by MHE, the modification of trajectories by MPC for the entire horizon could not be feasible by the lack of computational time. This forces MPC to solve the optimal control problem with a limited length of prediction horizon where the terminal state of the bioreactor is not considered. This becomes a problem because most of the important cost is imposed only at the terminal state of the bioreactor. Therefore, providing a proper terminal cost of MPC becomes a crucial point to properly modify the trajectories. At this point, the value calculated from the RL could be used as the terminal cost of MPC. Although this value would not be exact by the changes of parameter values, the simulation results suggest that this approximation is valid for MPC to stably calculate the new trajectories.

The key to this combination of MPC and RL is selecting the proper length of the prediction horizon. Under this scheme, the length of the prediction horizon is not only determined by the computational limits, but it is related to the reliability of the parameter values. The long prediction horizon implies that MPC is more rely on the newly estimated parameter values than the one used for RL. The presence of the parameter changes is one support on using the longer prediction horizon. However, the reliability of the estimated parameter values is related to both the quality and quantity of measurement data. Therefore, the choice of prediction horizon length should be carefully determined by evaluating the reliability of the parameter values. We propose one method for evaluating such reliability and propose a method of adjusting the prediction horizon length.

[1] C. Larroche, Sanromán Maria Ángeles, G. Du, and A. Pandey, Current developments in biotechnology and bioengineering. Amsterdam: Elsevier, 2017.

[2] Shuler and Kargi, Bioprocess engineering. United States: Academic Internet Publishers, 2007.

[3] H.-H. Kuo, Introduction to stochastic integration. New York: Springer, 2006.

[4] R. S. Sutton, Reinforcement learning: an introduction. Cambridge, Mass: MIT Press, 2018.

Copyright © American Institute of Chemical Engineers. All rights reserved.

608716 Sustainable API Development: Developing a Strategy for Measuring and Reducing Environmental Impact in the Pharmaceutical Industry

Abstract:

In January 2020, AstraZeneca announced its ambition to eliminate greenhouse gas emissions from its sites and fleet by 2025 and to become carbon negative across the entire value chain by 2030. The development and manufacture of active pharmaceutical ingredients (API) contributes significantly to AstraZeneca's carbon footprint and process engineers have a responsibility to play a leading role in achieving the zero-carbon target.

This presentation will share the approach taken to implement a strategy for measuring and reducing environmental impact during the development and manufacture of APIs. This approach comprises of influencing ways of working during API process development, implementing robust measurement of process environmental impact and identifying opportunities for innovation through science and technology to minimise AZ's carbon footprint.

This approach describes developing a procedure to appropriately consider sustainability alongside other key drivers for process design such as product quality, manufacturability and regulatory factors. Similarly, creating a culture to deliver continuous improvement against established metrics for sustainability in every aspect of API development and manufacturing.

This case also describes initiation of novel technology investigations to significantly reduce the carbon footprint of AstraZeneca's API manufacturing processes. This will be a leading factor in achieving a 50% reduction in carbon footprint for API development and the zero-carbon ambition of the company.

Copyright © American Institute of Chemical Engineers. All rights reserved.

609303 An Integrated Process Automation System in API Manufacturing: Key Characteristics and Case Study

Abstract:

Safety is always first priority and a critical concern in the chemical industry. Jiangsu Provincial Emergency Management Administration (JSEMA) in China released <Basic Requirements for Diagnosis and Improvement of Instinct Safety> (Su Emer. NO.53) in June 2019, which specifies the requirements for the main aspects of chemical production. The concept of "integrated process automation" in Su Emer. No. 53 poses a great challenge to API manufacturing. It emphasizes mandatory safety and intensified automatic implementation. This paper illustrates the key characteristics in the integrated process automation of API manufacturing and gives some case studies.

Such automation system includes sophisticated solution and cutting-edge tools. According to the recipe and process of the product, the consultant optimizes the control strategic with expertise to enhance the safety and meet the regulation requirements. The process control modules or batch control program are deployed in DCS system for the purpose of multifunctional and flexible manufacturing. In case study it gives example of the mixing control strategic in the reactors which is critical and should consider all concerned factors. Usually the geared motor electricity is ignored to monitor, the absence may interfere the cooling system activation in time, its real time monitoring and interlock with cooling media are necessary; another example is the utility facilities configuration and interlock with automatic systems which should be tuned duly to eliminate the hidden trouble. The integrity of safety instrumented functions (SIFs) should be determined through risk assessment tools and rating tools, such as HAZOP (Hazard and Operability Analysis) and LOPA (Layer of Protection Analysis). The related case study reports sharing control components between SIS and DCS will increase the possibility of failure, which should be validated. The allocation of center control room and controller assignment should be concerned with HSE guideline. It also discusses the roadmap of information technology and MES (Manufacture Execution System) for API manufacturing. Our implementation shows that the integrated process automation system can help to reach instinct safety, which increase the productivity and improve the quality of products with regulatory compliance.

Copyright © American Institute of Chemical Engineers. All rights reserved.

605455 A Quantitative Assay of Sodium Triacetoxyborohydride

Abstract:

Sodium triacetoxyborohydride (STAB) is a common reducing agent for reductive amination reactions. Due to sensitivities to water and air, the potency of this reagent tends to degrade over time. Traditional analytical assays for STAB fail to accurately assess the content of the active borohydride species, and often report assay purities of 97% or greater, when the true potency is lower. An incorrect or unknown potency of the STAB being used may lead to significant challenges when scaling up reductive aminations. An excess of STAB may be required and reactions may not achieve full conversion with the amount of STAB charged, requiring additional charges. The use of an excess of STAB comes with greater raw material costs and the thermal and H_2 evolution hazards associated with quenching the excess reagent.

This presentation outlines a simple assay, based on an aldehyde reduction, that was developed to determine the active borohydride content of STAB. The HPLC assay yield of a salicylaldehyde reduction has been shown to accurately determine this potency and has been validated against an H_2 evolution method, as well as yields obtained from a reductive amination.

Copyright © American Institute of Chemical Engineers. All rights reserved.

607708 Delivery Strategies for Live Therapeutic Bacteria

Abstract:

The microbiome is a collection of bacteria, viruses and fungi that perform essential functions for human health. Dysbiosis, or the dysregulation of the microbiome, can contribute to chronic inflammation, metabolic disorders, neurological diseases, and cancer. Clinically, one of the only methods for altering the composition of the microbiome is with broad-spectrum antibiotics, which contribute to dysbiosis and create selective pressure for antibiotic resistant pathogens. As such, live therapeutic bacteria (LTB) have been identified as an attractive alternative to antibiotics, as they can modulate the composition of the microbiome without eliminating commensal species. However, LTBs are commonly delivered orally and experience significant challenges during gastrointestinal (GI) transit, including survival in the harsh conditions of the stomach, adherence to the intestinal lumen, and navigation of interpatient variability in the microbiome due to factors such as diet. Few engineered drug delivery platforms have been rationally developed to address these challenges. In this work, we describe two complementary platforms that separately adapt LTBs to improve viability during gastric transit and modify LTB surfaces to aid in adhesion along the GI tract. In the first system, we sought to understand the influence of factors such as diet on LTB adaptation and colonization. We delivered an LTB to female BALB/c mice on two distinct diets and found that diet significantly influenced both the appearance of a small colony variant of the LTB, as well as its distribution along the GI tract. As adaptation appeared to significantly affect LTB colonization, we developed an approach to control the adaptation of LTBs to specific environmental insults during GI transit. We found that growth in an *in vitro* adaptation media can prime LTBs for the acidic conditions of the stomach, significantly improving their viability during a challenge in simulated gastric fluid. Following transit through the stomach, LTBs must adhere to the intestinal lumen to persist and colonize in the GI tract. Therefore, for our second system, we developed a platform to chemically conjugate targeting ligands to the surface of LTBs to improve their adhesion to the GI mucosa. This platform significantly improves both the rate of colonization, as well as the concentration of LTBs in the GI tract of female BALB/c mice. Collectively, this work represents a simple approach to adapt LTBs to the intestinal environment during *in vitro* growth, while enabling subsequent modification of the LTB surface for improved intestinal adherence. Together, these delivery techniques for LTBs may help overcome challenges associated with GI transit, colonization and adaptation.

Copyright © American Institute of Chemical Engineers. All rights reserved.

609030 In Vitro Biomimetic Tumor Model with Cancer-Specific Gene Regulators and Green Fluorescence Reporter for High Throughput Drug Screening

Abstract:

Although much scientific effort has been expended for cancer treatment, this disease is still one of the leading causes of death around the world. The limitations and shortcomings of current therapies highlight the urgent need for more effective therapeutic strategies. So far, most of the chemotherapy drugs act negatively on the cell division without selection, leading to serious adverse effect on normal cells. The current mainstream of anti-cancer drug development is to find the drugs that have specific targets (DNA, mRNA or proteins). Those targets are overexpressed in cancerous cells and are critical for carcinogenesis. In this study, we engineered breast and lung cancer cells to express the enhanced green fluorescent protein (EGFP) under cancer-specific gene promoters for three known carcinogenesis genes - survivin, Bcl-2 and hTERT, which enable online high-throughput screening of drugs that may up- or down-regulate these cancer genes as indicative of efficacy in inhibiting breast and lung cancers. Additionally, the mechanism of drugs and the crosstalk between cancer genes can be elucidated at the early stage of drug discovery. Another issue in the pharmaceutical industry is the high drug failure rate owing to poor predictability of early drug assays in a two-dimensional cell culture system. In such traditional assay, multiple cell-cell and cell-matrix interactions are deficient, leading to incomprehensive cellar responses to drug agents. To improve the drug assay predictability, our assay has focused on the construction of three-dimensional biomimetic cancer cell models with the recapitulation of multiple elements of tumor microenvironment. While developing the new drug screening assay, we also combined the rapid cell detection approach to achieve the high-throughput capability. Our optimized drug screening process can help to discover more anti-cancer therapies.

Copyright © American Institute of Chemical Engineers. All rights reserved.

609939 Preparation and Use of Fine Grade Engineered Excipients for Direct Compression of Binary Blends of Cohesive Drug Powders

Abstract:

Novel, surface modified excipients are prepared using a fluid energy mill (FEM) to simultaneously mill and dry coat microcrystalline cellulose (MCC) with nano-silica to produce different sizes of fine, well flowing MCC excipient [1]. The purpose of these new class of excipients is to enable direct compression (DC). DC route is one of the most efficient and economic methods for tablet manufacturing since only blending and compression operations are required, greatly simplifies the process-train. However, the popularity of direct compression is limited by its stringent requirements, such as good flowability, high packing density, and excellent compaction for excipient-active pharmaceutical ingredient (API) blends. When the blend has high drug loading with a poorly flowing API, the flowability and tabletability of blends is generally decreased sharply. In a recent work, it has been shown that surprisingly, fine excipients of 20 mm size that have improved flow and bulk density after silica dry coating, can greatly improve direct compression performance [2]. Unfortunately, it could not work for very fine cohesive APIs such as 10 mm acetaminophen (APAP) at higher drug loading. In this work, binary blends of several fine sized dry coated MCC-based excipients are considered with three different fine, cohesive APIs at high drug loadings. Bulk density, FFC, and tablet compactability for all these blends are measured and compared to determine which excipient sizes provide the best overall performance. The results are analyzed for the relative sizes of the excipients and APIs to better understand the interplay between the components of the blends and assessment of the extent of silica transfer from coated excipient powders to API powders and its effect on blend bulk density and flowability. It is expected that using such novel surface engineered fine excipients in blends with fine APIs may also help in reduced segregation tendency during overall processing. Hence, these results are expected to provide industry relevant guidelines for the use of finer surface modified excipients in DC tableting.

Copyright © American Institute of Chemical Engineers. All rights reserved.

596229 Applying Machine Learning to Predict Therapeutic Antibody Viscosity

Abstract:

A general machine learning feature selection protocol to predict antibody stabilities is proposed. These features only require information from sequences and molecular structures of antibodies. In addition, the physically relevant features are easily interpretable and help to design experiments to understand the underlying mechanisms of antibody stabilities. Two case studies are demonstrating using this protocol. First, we applied this framework to predict aggregation behaviors of 21 marketed monoclonal antibodies at high concentration (150 mg/mL), yielding an accuracy of 97% on validation tests with only three features. Models trained using structural features, rather than sequence features, correlated better with aggregation. Second, the concentration dependent viscosity behavior of 27 FDA approved antibody drugs was measured in concentrated solutions at high concentration (150 mg/mL). Combining molecular modeling and machine learning feature selection, we found that the net charge in the antibodies and the amino acid composition in the variable region are key features which govern the viscosity behavior. We presented two predictive models for aggregation and viscosity based on machine learning trained models, facilitating early stage design.

Copyright © American Institute of Chemical Engineers. All rights reserved.

602107 Development of Multi-Adjuvant Systems for Vaccines Against Infectious Diseases

Abstract:

The ability to elicit potent humoral and cellular immune responses is a key element required for the development of efficacious prophylactic and therapeutic vaccine formulations. For vaccines against intracellular pathogens, the induction of a balanced, long-lived, and potent $CD4^+$ and $CD8^+$ T cell activation and proliferation is critical, combined with robust antibody responses. One strategy is to activate innate immunity and proinflammatory responses through Toll-like receptor-independent cytosolic pathways. Importantly, engagement of these mechanisms has demonstrated its ability to provide protection against bacterial challenge, and several small molecules have been evaluated in preclinical and clinical studies as vaccine adjuvants against infectious diseases.

Small molecules can bind to cytoplasmic receptors triggering signaling cascades that activate innate immunity. Despite their potential as adjuvants, small molecules have some limitations that need to be overcome. For example, there may be low exposure in target tissues due to their size and charge, and are prone to enzymatic biodegradation; with some off-target toxicities. In order to address these challenges, the use of drug delivery systems is being explored.

In this work, three different polymeric nanoparticle platforms were evaluated for small molecule delivery and for their capability to act synergistically as co-adjuvants and elicit robust immune responses. In these studies, nanoparticles based on natural and synthetic polymers were fabricated using water/oil/water and oil/water emulsions. Particle characterization was performed using dynamic light scattering, scanning electron microscopy, and zeta potential measurements that confirmed the size, morphology, and charge of the resulting nanoparticles. Small molecule compounds were loaded into each of the platforms, and their loading and release efficiency were determined using liquid chromatography. Finally, these systems were evaluated using *in vivo* murine models following parenteral administration to assess their ability to induce cytokine secretion, antibody and cell-mediated immunity. The results obtained demonstrate the potential of polymeric particulate systems as dual delivery and co-adjuvant platforms for the development of small molecule-adjuvanted prophylactic and therapeutic vaccines.

Copyright © American Institute of Chemical Engineers. All rights reserved.

603577 Development of a Harvest Small Scale Model to Simulate Continuous Centrifugation

Abstract:

Disk stack centrifugation is one of the main clarification steps used in large scale mammalian cell culture processes. During process development, pilot scale disc stack centrifuge is normally used to develop and characterize the harvest process for new cell culture processes due to the lack of smaller scale model for disc stack centrifugation. In order to increase speed of development, a small scale model (SSM) of a disk stack centrifuge is of high demand to enable harvest development in an earlier stage of process development. A small scale model can also improve throughput and reduce cost. In this study, we have developed a SSM for disc stack centrifugation using a capillary shear device combined with benchtop centrifugation. Capillary tube size, length, and flow rate were optimized to generate energy dissipation rates comparable to large scale disk stack centrifuges. Additionally, shear force from multiple centrifuges at different scales were standardized and compared. Mathematical correlations were established for energy dissipation, viable cell density, cell viability, and lactate dehydrogenase (LDH) activity. Our SSM offers an approach to understand how the separation process, such as centrifuge designs, sizes, and configurations affect filtration performance and product quality. This work can speed up harvest development during process scale-up in biopharmaceutical mammalian cell culture processes.

Copyright © American Institute of Chemical Engineers. All rights reserved.

599615 Optimizing and Reducing Cleaning Production Loss in Pharma, Agrochemical and Specialty Chemical Companies Using a New Expert System.

Abstract:

Pharmaceutical, Biopharmaceutical, Agrochemical and Specialty chemical companies make many different products in their plants. This is true both for the active ingredient stages as well as the secondary of formulation processing. Cleaning and set up between products is critical and reduce productivity time up to 50% in some cases. The presentation will show the critical areas important in the modelling development in this expert system and how it will aid companies in scheduling their productions to reduce cleaning time and to reduce issues with cleaning.

Copyright © American Institute of Chemical Engineers. All rights reserved.

599656 Revolutionize of Pharmaceutical Product Changeovers By Enhanced Cleaning with Microbubbles

Abstract:

Pharmaceutical, Biopharmaceutical, Agrochemical and Specialty chemical companies make many different products in their plants. This is true both for the active ingredient stages as well as the secondary of formulation processing. Cleaning and set up between products is critical and reduce productivity time up to 50% in some cases. Microbubbles have been shown to enhance many unit operations e.g. distillation. The benefits of microbubbles in cleaning will be presented along with the impact this will enable in regulatory/ registration processes.

Copyright © American Institute of Chemical Engineers. All rights reserved.

611628 Gas-Phase Cyclic Voltammetry for Detection of Tuberculosis Associated Volatile Organic Compounds in Breath Using Co-Tna

Abstract:

Tuberculosis (TB) is one of the top 10 causes of death worldwide, leading to an estimated 1.5 million deaths in 2019. The World Health Organization identifies the development of rapid diagnostics at the point of care (POC) for TB as high priority [1]. Their annual report indicates that when diagnosed early, the majority of TB positive patients can be cured by first-line antibiotics. However, traditional detection methods such as sputum culture have incubation times of days or weeks, and require expensive instrumentation, laboratory level spaces and highly trained technicians. The need for repeated visits to medical facilities for sample delivery and test results is prohibitive to many, and results in significant patient dropout. In addition, up to one third of adults and most children are unable to produce sputum for traditional TB diagnostic techniques, driving an increased need for rapid non-sputum based testing. It is imperative to develop these POC tests at low cost, as over 95% of cases occur in countries classified as low or middle income. Screening of TB by evaluation of associated volatile organic compounds (VOCs) in breath has been shown to be a promising technology that is significantly faster and more convenient than traditional sputum-based tests. Methyl nicotinate (MN) and methyl p-anisate (MPA) have been identified as breath biomarkers specific to mycobacterium tuberculosis and are not found readily in the breath of healthy patients [2], making them ideal candidates for the diagnosis of TB in breath.

MN and MPA can be electrochemically detected using a cobalt-functionalized titanium dioxide nanotube array (Co-TNA) sensor, as shown by Bhattacharyya et. al [3]. Specificity is attained by the functional metal (in this case cobalt) interacting with the analytes of interest reproducibly. Traditionally, this sensing platform was operated using a simple two-electrode setup and chronoamperometric methods, in which a constant bias voltage is applied to the sensor surface and analyte interactions are observed by the resulting current change. However, chronoamperometric techniques are prone to confounders, as they are not able to separate signals from interactions occurring at smaller potentials or those resulting from surface resistance changes, for example when excess water is present. This is especially problematic in human breath which can contain hundreds of unique VOCs. In liquid systems, the presence of a third reference electrode allows this problem to be mitigated by employing scanning potential techniques, which allow the analyst to observe interactions at many potentials rather than just one. Therefore, our lab has modified the sensing platform to use a 3D-printed electron-conductive polymer as a solid-state electrolyte analogous to a salt solution in traditional liquid electrochemistry in order to create a gas-phase three-electrode system. This modification means that complex potential-scanning techniques can be performed on a fully solid-state sensor, a unique ability that is extremely important when running tests at the point-of-care where a long sensor shelf life is often necessary.

Using this new platform, cyclic voltammetry (CV) methods have shown promise in detecting TB VOCs more specifically than chronoamperometry alone. Features of interest present in CVs such as peaks, hysteresis, gradients, adsorption/ desorption events and current magnitudes are indicative of specific sensor interactions. Examples of these features can be seen in Figure 1. Lab studies using as little as 1 L of breath result in significantly different CV curves for clean vs MN spiked breath (an important TB biomarker). Figure 2 illustrates these differences, in which the CV of MN spiked breath exhibits a variance from the traditional smooth CV curve in the region where MN is expected to interact with cobalt (< 0.5 V), as well as hysteresis on the back scan (a peak which causes the overlapping of the forward scan). Cataloging the presence of these features and the voltage potentials at which they occur over a large set of breath profiles will allow us to understand how the sensor responds differently when exposed the breath of TB positive vs negative patients. This is imperative to developing a useful TB diagnostic tool. Therefore we utilize numerical methods via Python in order to quantify results from the obtained CVs. Average and localized gradients, peak detection, numerical integration for current magnitude, and hysteresis detection via regression analysis and optimization methods will be used to quantify data and make conclusions.

In order to better understand how these sensors perform on suspected TB patients, an ongoing field study is being conducted in collaboration with Makerere University located in Kampala, Uganda. Breath is collected from patients in Tedlar® bags and tested using a setup identical to that used in the lab. The results are analyzed using the numerical methods described, creating a catalog of interactions seen in true patient breath. The patients then undergo standard tuberculosis testing via collected sputum using smear microscopy, GeneXpert® testing and mycobacterial culture to confirm TB status. This allows us to better refine our parameters for classifying TB positive and negative patients, increasing the specificity of the sensor. These results are then validated by testing for the presence of TB biomarkers using thermal desorption coupled with GC-MS. Preliminary results shown in figure 3 illustrate the differences between CVs obtained from TB positive and negative patients. The resulting curves show a similar pattern to those obtained in lab mimics with healthy vs MN spiked breath (see figure 2). Once a significant number of patients have been tested, computer-driven analysis of patient curves can be used to develop accurate correlations in order to distinguish between TB positive and negative patients.

Data comparing sensor and GC-MS results obtained from lab mimics as well as confirmed TB patients will be presented.

References:

[1] World Health Organization, "Status of the TB epidemic," 2019.

[2] Syhre M, Chambers S, "The scent of Mycobacterium tuberculosis." Tuberculosis, 2008; 88:317-23.

[3] D. Bhattacharyya, Y. R. Smith, S. K. Mohanty, and M. Misra, "Titania Nanotube Array Sensor for Electrochemical Detection of Four Predominate Tuberculosis Volatile Biomarkers," J. Electrochem. Soc., vol. 163, no. 6, pp. B206–B214, 2016.

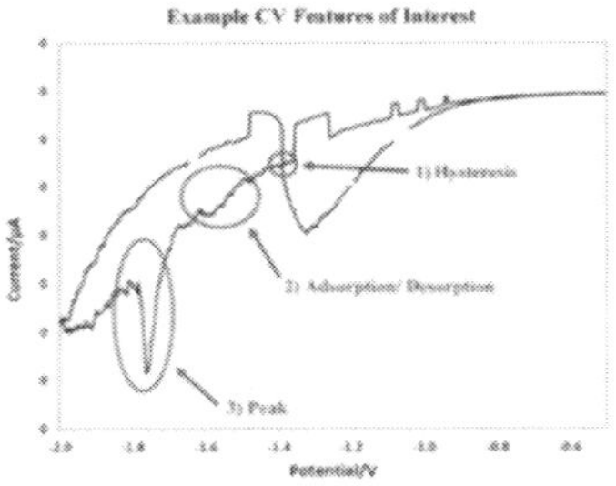

Figure 1. Example of CV containing features indicative of interactions with the sensor: 1) Hysteresis, occurring when the presence of a peak causes the forward and backward scans to cross; 2) Adsorption/ desorption event at the electrode surface, characterized by the rapid oscillation in the scan; 3) Peak, indicative of a diffusion limited reaction. Other CV characteristics to be catalogues include the gradient of the drops and the change in CV with scan number

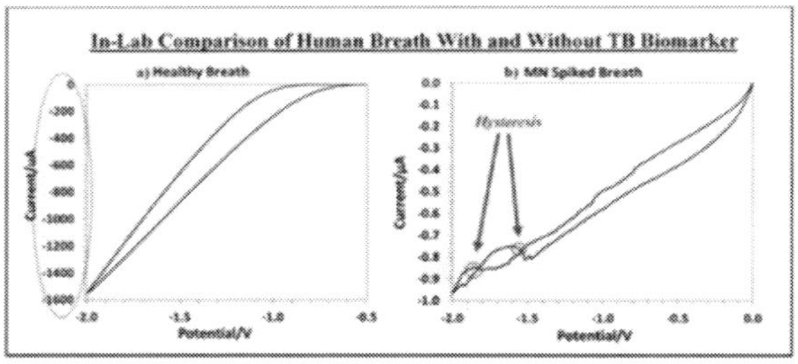

Figure 2. **a)** CV of a sample of breath from a healthy analyst using Co-TNA sensor, **b)** CV of a sample of breath from the same analyst spiked with TB biomarker MN using Co-TNA sensor. Variances from smooth CV curve seen in **a)** are labelled, such as peaks and max current magnitude

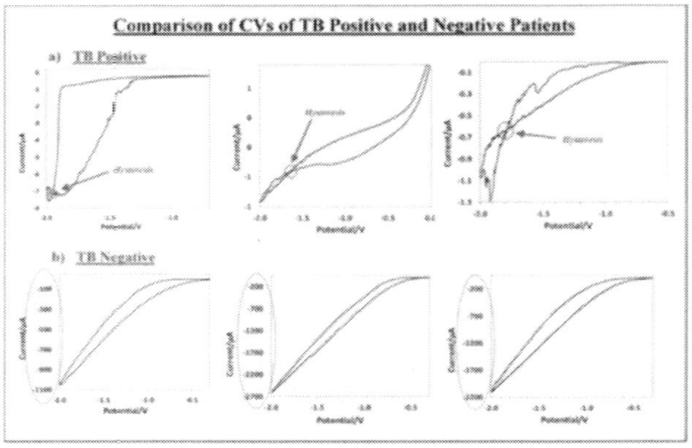

Figure 3. Preliminary sensor results from suspected TB patients in Kampala, Uganda: **a)** confirmed positive for TB via GeneXpert®, or **b)** confirmed negative via sputum culture. Positive patients often show curves with hysteresis at low potentials and max current magnitudes between 1-10 uA. Negative patients do not exhibit hysteresis and generally have max current magnitudes over 1000 uA, aligning with results obtained in lab studies.

Copyright © American Institute of Chemical Engineers. All rights reserved.

602368 Bacterial Separation from Blood through Sedimentational Clotting

Abstract:

Rapid diagnosis of blood infections presents many challenges, the first of which is fast and efficient separation of the bacteria from the blood cells. We previously developed a spinning hollow disk that separates bacteria from blood cells via the differences in sedimentation velocities of these particles. Blood was collected in EDTA tubes from volunteers and spiked with a known amount of E. coli bacteria; and then the separation of the bacteria from the blood was measured. Utilizing our spinning disk to separate the blood cells from the bacteria, we attempted to clot the red cell pack so that the plasma containing bacteria could be quickly collected without remixing with the blood cells. Clotting was induced by adding 8 M calcium chloride and adenosine diphosphate (ADP) to the blood on the disk. The CaCl2 was added to replace the calcium chelated by EDTA in the blood collection tubes. The ADP was added to induce the platelets to aggregate. Factors affecting the bacterial separation included the amount of adenosine diphosphate added to the blood, spinning parameters (speed and duration), and disk design. These factors were varied in dozens of experiments for which the concentration of bacteria in the separated plasma was measured. Results show that the best disk design contained a separation weir built into the disk, and the optimum spinning parameters were a separation sequence for 10 seconds followed by a clotting sequence for 62 s without the use of any ADP. These parameters achieved 69 ± 7% bacterial recovery from 7 mL of whole blood in 117 s. This is comparable in time and collection efficiency to previous separation of bacteria from blood using carefully designed weirs and controlled deceleration.

Copyright © American Institute of Chemical Engineers. All rights reserved.

608350 Automated mRNA Display for Rapid, Affordable, Target-Directed Reagent Development

Abstract:

As more potential cancer targets are being identified, there is a pressing demand for affinity reagents that can specifically bind to them. These reagents are essential to further research on the targets for better cancer diagnosis and treatments. The conventional reagent development methods are unreliable, slow, and expensive to perform. As a result, reagent development has been limited to large research facilities, greatly hindering our advancement in next-generation technologies to fight cancer. In light of these challenges, we set out to build a new reagent development pipeline through automating mRNA display. mRNA display is a powerful technology that screens a library with trillions of members for ultra-high affinity molecules. The target-directed reagents are developed through a cyclic in vitro evolution process that is typically repeated 5-15 times. By automating mRNA display, we aim to transform the technically challenging biochemical process into a broadly-accessible toolset.

To automate mRNA display, we converted the cyclic process into a network of functional microfluidic components using programmable equipment. The microfluidic components are fabricated using a bench-top 3D printer, which enables straightforward, rapid, and low-cost prototyping of the devices throughout the system development process. For instance, we designed a microfluidic enrichment device (MFED) for automating the bead-based purification steps in mRNA display. In bead-based purifications, peptides are allowed to bind the bead-immobilized target and the ones that are not bound to the target are removed. When performed manually, the removal of unbound peptides is done by repeatedly washing the beads using centrifugal filters. This approach, however, is not applicable to automation. Instead, we automated the purification steps using the MFED which immobilizes the target-bead complexes within a microfluidic channel. The device enables automated bead washing in a continuous-flow manner.

To benchmark the performance of the automated system, we performed controlled enrichments using previously prepared ligands against B-cell lymphoma-extra large (Bcl-x_L), a pro-survival cancer protein. We demonstrated that the MFED considerably outperforms the manual technique, with higher enrichment capability per round of selection. Using the MFED, the automated system can further accelerate mRNA display and lower the cost to generate reagents because of its superior performance in selection.

The automated mRNA display has transformative potential in addressing the current challenges of reagent development. By combining microfluidic automation and mRNA display, we aim to create an inexpensive platform for ligand discovery, expand the use of reagents in the biomedical and biopharma community, and thereby accelerate the evolution of cancer diagnostic and therapeutic tools.

Copyright © American Institute of Chemical Engineers. All rights reserved.

609239 Rapid Molecular Analysis of Human Tumor Samples in Point-of-Care Settings

Abstract:

One of the biggest cancer challenges in low- and middle-income countries (LMICs) is providing rapid and affordable diagnostics that enable patients to obtain locally available treatments. Unfortunately, in many LMICs, diagnosis often takes an extraordinarily long 6-12 months, largely due to lack of pathologists and limited infrastructure[1]. To address these barriers to diagnosis, we developed an integrated diagnostic platform with automated analyses, termed CytoPAN—portable fluorescence-based image cytometry analyzer—that can be used by non-expert healthcare providers[2]. We integrated three complementary approaches into a new 'all-in-one' platform: i) a disposable fluidic cartridge and prepackaged reagent kits for on-site sample processing, ii) a multi-channel image cytometry analyzer, and iii) a custom-designed software package for automated analysis. In our pilot clinical study with 68 prospectively collected patient samples, we validated CytoPAN's measurements against the gold standard histopathology; the accuracy for cancer detection was 100% ($P < 0.0001$, unpaired t test), and the accuracy for receptor subtyping was 97% for HER2 and 93% for ER/PR[2]. Unlike the conventional cytopathological diagnoses that requires trained pathologists and large number of cells, CytoPAN performs fast and accurate cancer diagnosis and molecular subtyping in *as few as 100 cells within an hour*. Based on these results, we envision prospective clinical trials in remote, decentralized locations. The system is also being deployed in extended field trials in Botswana, funded in part by an NIH global health initiative. The methods will be broadly applicable for other highly prevalent cancers affecting minority and underserved populations *domestically* and *abroad*. This platform may alter therapeutic paradigms for breast cancer patients globally and enable the appropriate use of chemotherapies and anti-hormones in limited supply.

REFERENCES

1. Nelson AM, et al. *J Clin Oncol*. 34(1) 20-26 (2016).

2. Min J, et al. *Sci Transl Med*. (2020, under re-revision)

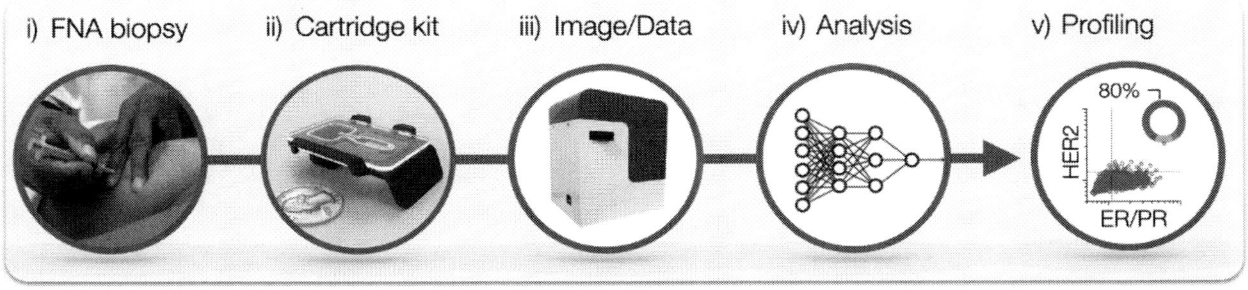

Copyright © American Institute of Chemical Engineers. All rights reserved.

610316 Polymeric Drug Delivery Systems for Small Hydrophilic Molecules

Abstract:

The clinical application of many drugs can be limited by rapid clearance from the systemic circulation or a local site of administration necessitating frequent dosing. This is particularly true of small molecule drugs of biopharmaceutical classification types I and III (hydrophilic or amphiphilic drugs). The pharmacokinetics and therefore the therapeutic effect of some such compounds can be improved by encapsulating them in delivery systems which allow sustained release, prolonging the duration of effect. However, the high-water solubility, very hydrophilic nature, and low molecular weight make it difficult to encapsulate small hydrophilic molecules in many conventional drug carriers, such as poly(lactic-co-glycolic acid) (PLGA) particles and hydrogels. Here, tetrodotoxin (TTX), a naturally occurring neurotoxin, was selected as a small hydrophilic molecule model. We will introduce some novel polymeric drug delivery systems that have been validated to efficiently encapsulate TTX and achieve TTX controlled release. Applications of these formulations in chronic pain management will be presented. The principles of polymer design and the mechanisms underlying the material performances will be discussed.

Copyright © American Institute of Chemical Engineers. All rights reserved.

611027 An All-Synthetic Platform for Highly-Targeted Immune-Activating Antibiotics Against Resistant Bacteria

Abstract:

The Centers for Disease Control and Prevention anticipates that antibiotic resistant pathogens will claim 10 million lives annually by 2050. This dire projection is driven by the lack of new antibiotics, which have moreover become challenging to discover by conventional techniques. We have developed a versatile new class of narrow-spectrum antibiotics, called immune-activating protein-catalyzed capture agents (IA-PCCs), that can be rapidly developed against designated resistant bacterial pathogens. IA-PCCs are bispecific compounds that comprise of an epitope-specific peptide ligand (the PCC) that binds the bacterium, and a hapten moiety that promotes immune clearance of the pathogen. Demonstrated here is the development of IA-PCCs against two resistant bacteria of high clinical significance: carbapenem-resistant *Klebsiella pneumoniae* and *methicillin-resistant Staphylococcus aureus* (MRSA). Integrated bioinformatics and multi-omic data analyses guided the judicious selection of abundant pathogen-specific epitope targets on the surface of each bacterium. Subsequent PCC screens against the selected epitopes yielded several macrocyclic ligands that exhibited high avidity to their molecular targets as well as whole *K. pneumoniae* and *MRSA* bacterial cells, as determined by enzyme-linked immunosorbent assays. Top-performing PCCs were conjugated with an antibody-recruiting hapten, dinitrophenyl (DNP), to form an IA-PCC. This IA-PCC promoted the recruitment of antibodies to bacterial surfaces and enhanced opsonophagocytic killing of resistant pathogen by murine bone-marrow derived macrophages. Variations on the immune-activating moiety are explored, by replacing the antibody-recruiting DNP label with molecular tags that directly activate immune cells, including formylated peptides, toll-like receptor agonists, and synthetic peptides. Lead IA-PCCs are subjected to medicinal-chemistry type iterations to improve physicochemical properties associated with *in vivo* performance, ultimately arriving at compounds ready for translation *in vivo*. *In vivo* mouse models to test IA-PCC efficacy are ongoing. The IA-PCC platform has emerged as a promising and general platform to address antibiotic resistance.

Copyright © American Institute of Chemical Engineers. All rights reserved.

611636 A Systems-Level Hypothesis for How Innate Immune System Dysregulation May Cause Alzheimer's Dementia, and How We May be Able to Prevent It

Abstract:

We are investigating the early-stage etiology of sporadic Alzheimer's Dementia (AD), for which 420+ clinical trials by Pharma have failed over the past 15 years to produce any effective drug. What causes the accumulation of Aβ peptide-rich fibrils and plaques in an aging brain? What is Aβ's physiological function? We focus on Aβ's interactions with the human cathelicidin peptide, LL-37, an antibacterial and antiviral innate immune system effector and modulator that is ubiquitous in human tissues and expressed by myriad cell types, yet unique in the human proteome. Recently, evidence has built that chronic *Herpesvirus* or *P. gingivalis* infections of human brain tissue may precipitate many cases of sporadic dementia labeled as Alzheimer's Disease. We present experimental evidence and discuss our developing hypothesis that the antiviral and antibacterial peptide LL-37, which can be chronically underexpressed in humans based on dietary and lifestyle factors or degraded by *P. gingivalis* virulence factors, is a natural binding partner of Aβ that inhibits formation of AD fibrils and plaques, such that LL-37 and Aβ have a toxin/antitoxin relationship. We demonstrate sequence-specific binding between LL-37 and Aβ by Transmission Electron Microscopy (TEM), Surface Plasmon Resonance Imaging (SPRi), circular dichroism (CD) spectroscopy, analytical ultracentrifugation (AUC) and Small-Angle X-Ray Scattering (SAXS). TEM shows that LL-37 inhibits the fibrillization of Aβ, especially the formation of long, straight fibrils characteristic of AD, while CD spectroscopy reveals that LL-37 binding prevents Aβ from adopting β-type secondary structure. AUC and SAXS show that LL-37 and Aβ form a unique, monodisperse, 1:1 molecular complex. Cell culture studies with primary human microglia and neuronal cells indicates that these two peptides, once they bind and form a complex, then neutralize each other's effects on the cells. Finally, studies in 5XFAD and wildtype transgenic mice, as well as the fruit fly *Drosophila Melanogaster*, support these findings. We discuss what all of this means in the context of the prevention and treatment of Alzheimer's dementia, and suggest several approaches that could be taken, in light of these results, to efficiently prevent (and possibly treat) AD.

Copyright © American Institute of Chemical Engineers. All rights reserved.

608455 Multi-"Color" Delineation of Bone Microdamages Using Ligand-Directed Sub-5 Nm Hafnia Nanodots and Photon Counting CT Imaging

Abstract:

The early detection of bone microdamages is crucial to make informed decisions about the therapy and taking precautionary treatments to avoid catastrophic fractures. Conventional computed tomography (CT) imaging faces obstacles in detecting bone microdamages due to the strong self-attenuation of photons from bone and poor spatial resolution. Recent advances in CT technology as well as novel imaging probes can address this problem effectively. Herein, the bone microdamage imaging is demonstrated using ligand-directed nanoparticles in conjunction with photon counting spectral CT. For the first time, Gram-scale synthesis of hafnia (HfO_2) nanoparticles is reported with surface modification by a chelator moiety. The feasibility of delineating these nanoparticles from bone and soft tissue of muscle is demonstrated with photon counting spectral CT equipped with advanced detector technology. The ex vivo and in vivo studies point to the accumulation of hafnia nanoparticles at microdamage site featuring distinct spectral signal. Due to their small sub-5 nm size, hafnia nanoparticles are excreted through reticuloendothelial system organs without noticeable aggregation while not triggering any adverse side effects based on histological and liver enzyme function assessments. These preclinical studies highlight the potential of HfO_2-based nanoparticle contrast agents for skeletal system diseases due to their well-placed K-edge binding energy.

Copyright © American Institute of Chemical Engineers. All rights reserved.

603892 (Invited) Skin-like Wearable and Implantable Sensors

Abstract:

In this talk, I will present our recent work on BodyNet concept for stretchable wearable and implantable sensors. Examples include, full body sensing, blood flow sensor and morphing electronics - neurostimulator.

Copyright © American Institute of Chemical Engineers. All rights reserved.

600113 An Adhesion-Based Anti-Corrosion Strategy for Wearable Electrochemical Sensing and System Integration

Abstract:

The large-scale deployment of biomonitoring sensors through integration in wearable consumer electronics enables population-level health and wellness monitoring, and thus can play a critical role in transforming personalized medicine. Currently, physical sensors have been widely incorporated within commercialized wearable platforms, but they cannot capture biomarker information that lie in molecular levels and provide highly specific information about the body's dynamic chemistry. Electrochemical biosensing interfaces can be positioned to target biomarker molecules in a wearable format, as they can be constructed on miniaturized/flexible footprints/substrates and can convert (bio)chemical signals into electrical signals measurable by electronic readouts. The contact pads on the printed circuit board (PCB), which is a core component of wearable consumer electronics, can directly serve as electrodes and as a substrate for electrochemical sensors. However, we found that the commercialized PCB metal contact pads (*e.g.*, gold) were prone to corrosion and unsuitable for electrochemical modifications in solution environments. Additionally, electrochemical biosensors often are not reusable and need to be replaced frequently, while PCB units are not positioned for disposable applications.

To this end, we devise an adhesion-based anti-corrosion strategy for PCB-interfaced electrochemical sensing, which enables seamless sensor-system integration. The strategy is based on decoupling the sensing interface and PCB metal contact pads by an intermediary ultrathin (~50 μm), vertical conductive, and double-sided adhesive film. This strategy has dual benefits of making electrochemical sensors disposable and insulating the contact pads from sounding solutions. We first establish the suitability of the adhesion film to serve as an electrode and as an interconnection film simultaneously. One side of the double-sided adhesive film was patterned with noble metal electrodes, followed by the modifications with recognition layers, such as permselective membrane and enzymes, to render biosensing functionality. The electrochemical biosensors were transferred and adhered to the PCB contact pads with the aid of water-soluble tape. As compared with bare PCB contact pads, this fabrication strategy allows for realizing a corrosion-resistant and stable electrochemical biosensing interface on PCB for operation in biofluid (*e.g.*, sweat). Amperometry and open circuit potential characterization experiments validated the electrochemical corrosion resistance properties in saline solution environments. The excellent anti-corrosion attributes of the modified film could be ascribed to: (1) the polymer glues on both sides of the adhesive film acting as corrosion-protective layers; (2) bypassing the need for oxidation prone-adhesion metal (*e.g.*, chromium), achieved by e-beam evaporation (at the noble metal patterning step); and (3) the electrodeposited permselective membrane (here, phenylenediamine), which has been reported to presented anti-corrosion properties. We specifically tailored the enzymatic layer of the electrochemical interface to target biomarker molecules such as glucose, lactate, and choline as example of informative metabolites and nutrients. By modifying different (bio)chemical layers various other molecules can be targeted. In this way, our strategy serves as a foundation for electrochemical sensor development and integration with wearable consumer electronics.

Copyright © American Institute of Chemical Engineers. All rights reserved.

607021 Rapid Screening Platform for the Isotopic Determination of Radionuclides in Water

Abstract:

Traditional approaches for the detection of trace-level radioisotopes in water require lengthy offsite sample preparations and do not lend themselves to rapid screening. Therefore, a novel sensor platform that combines onsite purification, concentration, and isotopic screening with a fieldable detection system will be an invaluable tool.

In this presentation, we will discuss findings on the development of a sensor platform based on reactive polymer film coatings for the rapid isotopic screening of waterborne radionuclides. Nanothin coatings were prepared by grafting uranium-selective polymers from polyethersulfone (PES) surfaces via UV-initiated polymerization, and by introducing uranium-selective functional groups to polyacrylonitrile (PAN) surfaces by chemical reaction. Ellipsometry was used to study film growth kinetics on PES films. X-ray photoelectron spectroscopy of modified PAN films revealed conversion of nitrile groups to amidoxime groups to be as high as 40% and showed that the extent and depth of reaction could be varied precisely. Static uptake experiments with solutions of depleted uranium spiked with ^{233}U were conducted to determine uranium binding capacities and kinetics of the modified polymer films at different pH values from 4 to 8. Sorption isotherm data were fitted to the Langmuir model, and the highest sorption capacities were obtained at pH 6 for modified PAN (M-PAN) and PES (M-PES) films. Capacities at pH 4 and 8 were lower and could be explained by differences in sorption mechanisms. Uranium batch uptake kinetics followed a pseudo-second order rate model. Alpha spectroscopy pulse height spectra were analyzed to study the role of selective layer film thickness on peak energy resolution. Excellent resolution with full width at half maximum values from 29 to 41 keV were recorded for M-PAN film and from 26 to 45 keV for M-PES film. Whereas uranium uptake increased with selective layer film thickness and varied with polymer chemistry/extent of modification, the peak energy resolution was independent of layer thickness and polymer chemistry within the experimental measurement uncertainties.

Results from this work were used to modify PES and PAN ultrafiltration membranes with the same experimental conditions to develop uranium selective membranes. High-throughput uptake experiments were performed to study the effect of residence time on detection efficiency and isotopic resolution. This membrane-based detection method enables the rapid, fieldable screening analysis of radionuclides in water for environmental monitoring studies and nuclear forensics investigations.

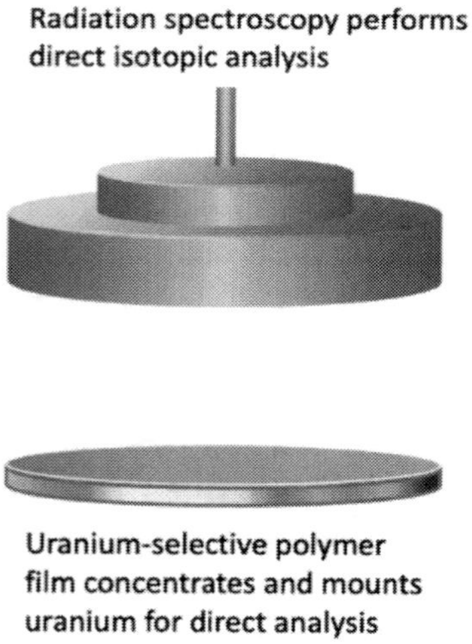

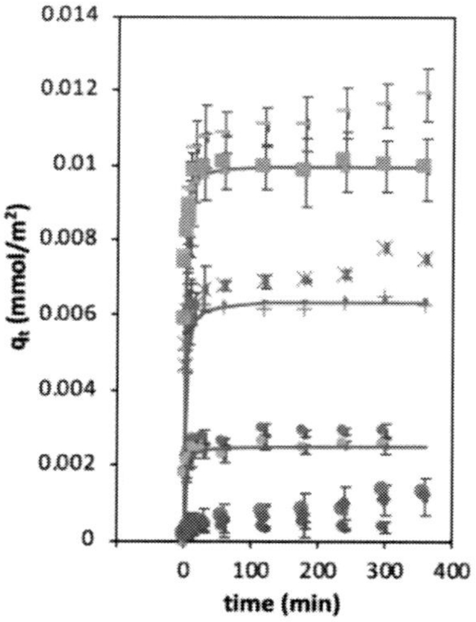

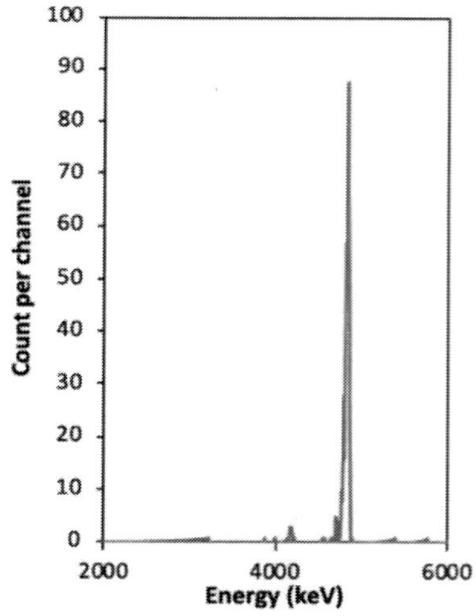

Copyright © American Institute of Chemical Engineers. All rights reserved.

600875 Electrochemical Detection of Environmental Pollutants

Abstract:

Many small molecule pollutants interact with human hormonal systems. These compounds, known as endocrine disruptors, are ubiquitous, as they are found in plastics, medications, and pesticides. Such chemicals are especially problematic because they have been linked to severe health problems, including cancer, infertility, and diabetes. We have developed two complementary electrochemical methods to rapidly detect and quantify these compounds. One technique involves the specific monitoring of Bisphenol A (BPA) using a DNA aptamer immobilized on an electrode. The second method involves the measurement of the biological activity of chemically dissimilar endocrine disruptors using a human hormone receptor as a component of an electrochemical sandwich assay. These electrochemical platforms require no specialized skills to implement and have enabled very sensitive detection (sub-ppb levels) of these environmental pollutants from complex solutions, such as baby formula.

Copyright © American Institute of Chemical Engineers. All rights reserved.

604144 Fabrication of High-Resolution Graphene-Based Flexible Electronics Via Polymer Casting and Microfluidic Approaches

Abstract:

Graphene-based flexible electronic devices for sensor applications has been receiving growing interest due to unique properties of graphene.[1] Various fabrication methods, such as photolithography, printing and chemical vapor deposition (CVD), has been used to produce high resolution and low feature size graphene-based devices.[2, 3] However, most of these methods are complex and require multiple processing steps (i.e. stamping, vacuum drying and etching) as well as additional post treatments (i.e. high temperature (300-1000 °C) baking or laser annealing) that can thermally or chemically degrade polymer-based substrates limiting substrate material selection.[2, 3] In addition, these methods can only be applied to 2D planar substrates and are not feasible to fabricate 3D circuits and some of them, such as sticky/adhesive tape peeling method, require high amounts of graphene consumption.[4] Despite the recent progress in the field, there is still a need to develop novel fabrication approaches that can address the current limitations of the existing methods.

In this study, we present two flexible electronics device fabrication methods based on simple polymer casting and microfluidics approaches.[5, 6] The first method focuses on simple polymer casting based graphene transfer at room temperature that does not require any post-processing. Briefly, our method consists of two main steps; (i) the formation of graphene patterns/films on substrates/molds via conventional methods such as CVD, channel filling or ink-jet printing and (ii) direct casting of target substrate polymer solution on the substrates/molds with graphene micropatterns and direct graphene transfer to the target substrate via peeling off upon drying and film formation. This method simply relies on the differences in the surface energies and adhesive forces between the graphene/mold and graphene/target polymer substrate.[5] Alternatively, the second method concentrates on a novel microfluidic approach to fabricate high resolution and low feature size graphene-based circuits/devices.[6] This method involves controlled and selective filling of microchannels on substrate surfaces with a conductive binder-free graphene nanoplatelets (GNP) solution. The ethanol-thermal reaction of GNP solution at low temperatures (~75 °C) prior to microchannel filling (as called "pre-heating") potentially reduces and homogenize GNP, which in turn enhances conductivity and reduces sheet resistance (down to ~0.05 kΩ sq^{-1}). Therefore, harsh post-processing of GNP is eliminated, and room temperature fabrication is enabled, which allows the use of degradable substrates, including biodegradable and natural polymers. This simple process is followed by selective administration of conductive GNP solution to the predetermined microfluidic channels on flexible substrates in a controlled manner via a syringe pump to create high resolution and low feature size graphene circuits/devices. This microfluidic approach can also be used to create 3D circuits with different geometries, using combined approaches with 3D printing, which is difficult to obtain with other methods (i.e. inkjet, screen or aerosol printing) that work mainly with 2D planar substrates. [6]

We demonstrated that using these fabrication methods, we obtained high resolution and low feature size graphene circuits with different geometries, dimensions and 3D superficial microstructures. We investigated the effect of temperature and microfluidic parameters (such as concentration, volume and flow rate) as well as the type of substrate materials on the graphene structure and formed micropatterns. The XPS, Raman, SEM and TEM results validated the structural changes of graphene along with confirming the stable presence of graphene patterns on the target substrates.[5, 6] The feature sizes of the graphene patterns could range from a few micrometers (down to ~15 μm in width and ~5 μm in depth) to a few millimeters with very small amounts of GNP used (~2.5 mg of graphene to obtain ~0.1 kΩ sq^{-1} of sheet resistance for 1 cm^2). [5, 6] The generated graphene patterns demonstrated significant stability after multiple washing and bending cycles. These methods were also implemented using other conductive inks, such as conductive graphene-silver based composite ink.[5, 6] Room temperature processing also provided precise control of 3D microstructural and mechanical properties (such as film porosity, pore size, elasticity etc.) of the target substrate materials, as demonstrated by SEM images.[5, 6] Using these methods, we successfully fabricated pressure sensors and near field communication antennas. [5, 6] Our results demonstrated that the devices worked with high sensitivity and stability.

In conclusion, the circuits/devices fabricated using this method can easily be implemented as wearable electronic, sensors and batteries, robotic components or motion detectors. In a broader sense, these simple, environmentally friendly and low-cost methods could pave the way for the fabrication of 2D and 3D electronic circuits/devices on different degradable flexible substrates using various graphene-based conductive inks.

References:

1. Jang, H.; Park, Y. J.; Chen, X.; Das, T.; Kim, M. S.; Ahn, J. H., Graphene-Based Flexible and Stretchable Electronics. *Advanced Materials* **2016,** *28* (22), 4184-4202.
2. Das, S. R.; Nian, Q.; Cargill, A. A.; Hondred, J. A.; Ding, S.; Saei, M.; Cheng, G. J.; Claussen, J. C., 3D nanostructured inkjet printed graphene via UV-pulsed laser irradiation enables paper-based electronics and electrochemical devices. *Nanoscale* **2016,** *8* (35), 15870-15879.
3. Das, S. R.; Srinivasan, S.; Stromberg, L. R.; He, Q.; Garland, N.; Straszheim, W. E.; Ajayan, P. M.; Balasubramanian, G.; Claussen, J. C., Superhydrophobic inkjet printed flexible graphene circuits via direct-pulsed laser writing. *Nanoscale* **2017,** *9* (48), 19058-19065.
4. Oren, S.; Ceylan, H.; Schnable Patrick, S.; Dong, L., High-Resolution Patterning and Transferring of Graphene-Based Nanomaterials onto Tape toward Roll-to-Roll Production of Tape-Based Wearable Sensors. *Advanced Materials Technologies* **2017,** *2* (12), 1700223.
5. Uz, M.; Jackson, K.; Donta, M. S.; Jung, J.; Lentner, M. T.; Hondred, J. A.; Claussen, J. C.; Mallapragada, S. K., Fabrication of High-resolution Graphene-based Flexible Electronics via Polymer Casting. *Scientific Reports* **2019,** *9* (1), 10595.
6. Uz, M.; Lentner, M. T.; Jackson, K.; Donta, M. S.; Jung, J.; Hondred, J.; Mach, E.; Claussen, J.; Mallapragada, S. K., Fabrication of Two-Dimensional and Three-Dimensional High-Resolution Binder-Free Graphene Circuits Using a Microfluidic Approach for Sensor Applications. *ACS Applied Materials & Interfaces* **2020,** *12* (11), 13529-13539.

Copyright © American Institute of Chemical Engineers. All rights reserved.

602434 A Biophysical Outlook of Viral Inactivation with Arginine and Common Buffers in Biologics Manufacturing

Abstract:

A need for robust virus inactivation processes and environmentally friendly materials had prompted an exploration of innovative and alternate strategies in biologics manufacturing. A common detergent, Triton X-100, often relied on for inactivation of enveloped viruses, is on the edge of disbarment due to its detrimental effect on the environment. In this interest, two inactivation methods based on arginine and commonly used buffers were explored. Past studies have shown arginine to inactivate certain viruses under various conditions. However, the optimal conditions and mechanisms for fully inactivating all enveloped viruses by arginine is not known. For this reason, we tested a matrix of parameters such as pH, incubation time and buffer type to discern an optimal condition for inactivation of all enveloped viruses. Our results demonstrated that larger viruses - herpes simplex virus (HSV-1) and pseudorabies virus (SuHV-1) were effectively inactivated with > 4 LRV as compared to the smaller sized viruses - bovine viral diarrhea virus (BVDV) and equine arteritis virus (EAV). However, the optimal conditions resulted in different inactivation values between the similar herpes family viruses, HSV-1 and SuHV-1. Mechanistic understanding is being derived from the study of mechanical properties, virus-host interaction, and changes in the virus membrane. The second study explored the efficiency of commonly used buffers in the downstream train. Various buffers such as glycine, acetate, and citrate at pH 4 were explored to determine the inactivation efficacy of the above-mentioned viruses. A similar trend of inactivation based on virus size was observed for all the buffers. The efficiency of the buffers considering both the composition and incubation time was in the order of glycine < acetate ≤ citrate. Overall, this work directs towards the trend in phasing out Triton X-100 inactivation of enveloped viruses during biologics manufacturing and effects of commonly used buffers in the biologic and viral DSP.

Copyright © American Institute of Chemical Engineers. All rights reserved.

611545 The Effects of Filtration Flux on the Clearance of Minute Virus of Mice during Viral Filtration

Abstract:

Viral safety is one of the major concerns in the production of mammalian cell and plasma derived biotherapeutic products. Demonstration of virus clearance is required by the regulatory agencies. Virus filtration is a size exclusion-based virus clearance method and is conducted industrially in dead-end mode. It is often operated in a batch mode under constant pressure. Membrane performance is shown to be affected by feed condition (pH, ionic strength and feed buffer components) as well as the operation condition (flux, pressure, time). During continuous bioprocessing, constant flux is a more desirable operation mode. Here the effects of flux operated under constant flux filtration mode and the solution condition on virus retention are systematically investigated for the filtration of monoclonal antibodies spiked with minute virus of mice (MVM) using three commercially available virus filters. Surprisingly, virus breakthrough was observed for two of the three filters when low constant flux operation was performed for a relatively longer periods of time. The interplay between virus loading on the filter, flux, membrane fouling and filtration time on virus retention will be discussed.

Copyright © American Institute of Chemical Engineers. All rights reserved.

611537 Outside-in Hemodialysis: A Study of Toxin Clearance

Abstract:

For a patient with chronic or acute kidney failure, it is advantageous to perform hemodialysis continuously and for long durations in order to gradually remove excess fluids and improve toxin removal. Unfortunately, such long-term treatments are often hampered by the short lifetimes of the hollow-fiber dialyzers used in toxin removal, which arise due to blood clots formed in these fibers. The blood clots often block an entire fiber, so as more fibers are clogged, the surface area of the dialyzer and waste clearance are drastically lowered.

Labib and Dukhin (Novaflux Inc.) postulated that this clotting limitation can be overcome by a new approach that switches flow on the blood and dialysate side. In the outside-in filtration (OIF) configuration, dialysate flows inside the fibers and blood flows through the inter-fiber space. The researchers showed that OIF would allow prolonged hemodiafiltration with little dialyzer clogging.[1] With OIF, even if small inlet thrombi form that would block fibers in conventional intraluminal blood flow, these clots will have minimal effect on the blood flow and filtrate flux in OIF due to the three-dimensional system of interconnected hydrodynamic flow channels in the inter fiber space. However, very few studies have looked into the impact of OIF on toxin clearance and the level of mass transport of solutes with different sizes. The objective of this study is to experimentally assess the transport of different types of uremic toxins under both outside-in (OIF) and inside-out (standard) configurations. For small solutes, toxin clearance is relatively similar between the OIF configuration and the standard configuration, for both diffusive and convective transport. For larger toxins, however, we observed a noticeable difference in clearance level when OIF mode is used. We will discuss the physical mechanisms behind these observations, and the consequences of these observations for clinical applications.

[1] Dukhin SS, Labib M, et al. Outside-in hemofiltration for prolonged operation without clogging. Journal of Membrane Science 464 (2014).

Copyright © American Institute of Chemical Engineers. All rights reserved.

600762 Study of the Spike (S) Glycoprotein from the Sars-Cov-2 As a Possible Source of Translocating Peptides of Biomedical Interest

Abstract:

Membranes are natural selective barriers necessary for the correct function of human cells as they control the entry of diverse molecules into the intracellular space. Although small and polar molecules such as amino acids and ions can come across the membrane through channels, larger macromolecules such as proteins and nucleic acids generally fail to do so. This represents a significant obstacle for efficient intracellular drug delivery, which is essential to increase the bioavailability of pharmacological agents for the treatment of numerous diseases[1,2]. Commonly, diverse methods have been studied to mediate the uptake of large molecules, including mechanical and electrical transfection techniques, but also viral carrier systems[2].

Some of these methods have been successfully tested *in vitro* but exhibit biosafety and cytotoxicity issues[3]. In consequence, there is an increasing interest in the development of novel delivery methods with low cytotoxicity but also high transduction efficiency[1]. An attractive alternative is the cell-penetrating peptides (CPPs), which can be obtained from different origins including signal peptides, viral proteins, or antimicrobial peptides[4]. CPPs are generally short peptides with lengths between 5-30 amino acids, positively charged or amphipathic, and rich in arginine and lysine[3]. Due to their ability to intermingle with the phospholipids of membrane bilayers, some CPPs are able to undergo translocation, and consequently can be considered as promising candidates for the delivery of biologically active molecules to cells.

The novel coronavirus SARS-CoV-2 has attracted significant attention over the past few months as it is responsible for the current global sanitary emergency where more than 1.3 M cases have been confirmed and over 70,000 people have died[5]. The Spike (S) glycoprotein has been thought to be responsible as the major determinant of the viral tropism towards human cells[6]. This protein has a 180 kDa molecular weight and is displayed at the viral surface as a trimer composed of two major domains[7]. The first one is the S1, which contains the receptor-binding domain (RBD) responsible for mediating the receptor binding (Angiotensin-converting enzyme 2). The second one is the S2, which allows the membrane fusion through the exposure of a fusion protein that is activated by proteolytic cleavage in a site upstream (S2') and proteolytically primed at the interface of the S1 and S2 domains. Transmission of the genetic material into the host cells has been attributed to proteases in priming, receptor binding, and some ionic interactions controlling the stability of the virus[7].

By recognizing the strong interaction between de spike (S) glycoprotein of SARS-CoV-2 and the angiotensin-converting enzyme 2 from the cellular membrane of the lung cells, here we aimed at finding motifs that could serve as possible sources of peptides capable of intermingling with membranes, and eventually with superior translocating potency. For this purpose, a prediction of the tertiary structure of the S protein from the SARS-CoV-2 was performed by homology, using Phyre Server, with the S protein from the bat coronavirus RaTG13. This was selected due to its closeness to the SARS-CoV-2 virus as it shares more than 93% of the identity in the S gene[8]. Simultaneously, the structure was also predicted *de novo* using the iTASSER Server to assure that predictions were robust enough for biophysical interaction studies. To determine the motifs with significant membrane activity (and potential translocation ability), a prediction of the possible formed transmembrane helices was carried out using TMHMM Server v. 2.0. Finally, the selected sequences were studied into detail via molecular dynamics (MD), using a mode membrane.

References:

4. **Derakhshankhah, and S. Jafari** "Cell penetrating peptides: A concise review with emphasis on biomedical applications" *Biomedicine & Pharmacotherapy*, 108, 1090–1096. (2018)
5. **W. Huang, and H.-J. Lee** "Cell-penetrating peptides for medical theranostics and targeted drug delivery" *Peptide Applications in Biomedicine, Biotechnology and Bioengineering*359–370. (2018)
6. **W. Huang, H.-J. Lee, L. M. Tolliver and R. S. Aronstam** "Delivery of Nucleic Acids and Nanomaterials by Cell-Penetrating Peptides: Opportunities and Challenges" *BioMed Research International*, 1–16. (2015)
7. **A. El-Baky, V. N. Uversky, and E. M. Redwan** "Virucidal activity of cell-penetrating peptides of viral origin"*Journal of Biomolecular Structure and Dynamics* 36, 1739–1746. (2017)
8. **H University and Medicine** "Coronavirus Resource Center" (2020)
9. **J.G. Hulswit, C.A.M. de Haan1 and B.-J. Bosch** "Coronavirus Spike Protein and Tropism Changes" Coronaviruses *Advances in Virus Research, 29–57.* (2016)
10. **A. Jaimes, N. M. André, J. K. Millet, and G. R. Whittaker** "Structural modeling of 2019-novel coronavirus (nCoV) spike protein reveals a proteolytically-sensitive activation loop as a distinguishing feature compared to SARS-CoV and related SARS-like coronaviruses" (2020)
11. **Lan, J. Ge, J. Yu, S. Shan, H. Zhou, S. Fan, Q. Zhang, X. Shi, Q. Wang, L. Zhang and X. Wang** "Structure of the SARS-CoV-2 spike receptor – binding domain bound to the ACE2 receptor" *Nature.* (2020)

Copyright © American Institute of Chemical Engineers. All rights reserved.

608800 Combinatorial Discovery of Ligand-Fucntionalized Microbeads for Selective Enrichment of Methylated and Unmethylated DNA

Abstract:

DNA methylation patterns play a vital role in determining gene expression in different diseases including cancer. In addition, therapeutic plasmid DNA (pDNA) molecules, used in non-viral gene therapy, are frequently methylated by the enzymes present in bacterial strains employed in their bio-processing. Current approaches to isolate methylated plasmids and genomic DNA involve antibodies, which are mainly geared towards research-scale applications. Consequently, new approaches that can lead to the enrichment of methylated or unmethylated DNA from biological samples and bioprocessing media, in a rapid and cost-effective manner, can lead to transformative advances in these fields. We have developed aminoglycoside-derived microbeads, approximately 10μm in diameter, for binding DNA based on Coulomb interactions between the beads and DNA molecules. These parental microbeads were modified with different amino acids, lipids, and chemotherapeutic drugs, in order to obtain a diverse library of ligand-modified microbeads. The library of ligand-modified microbeads was screened for binding unmethylated and methylated pDNA (a model plasmid that encodes for luciferase protein) in order to identify lead candidates that showed preferential binding to methylated pDNA over the unmethylated analog or vice versa. Adsorption isotherms were determined for both sets of leads and the data were fit to a Langmuir isotherm model. Elution of adsorpbed DNA was investigated with different buffers and organic modifiers. Finally, quantitative structure activity relationship (QSAR) models were employed to obtain physicochemical insights into molecular interactions that govern ligand binding to methylated as well as unmethylated DNA. In conclusion, our results indicate that molecular engineering of a chemically diverse ligand-modified microbead platform can lead to rapid identification of novel ligands for enriching methylated or unmethylated DNA for applications in bio-manufacturing of gene therapy vectors and in the rapid detection of diseases, including in low-resource settings.

Copyright © American Institute of Chemical Engineers. All rights reserved.

607410 Apohemoglobin Haptoglobin Complexes Attenuate the Hypertensive Response to Low Molecular Weight Polymerized Hemoglobin

Abstract:

Background: The presence of low molecular weight (MW) hemoglobin (Hb) species in previous generations of Hb-based oxygen (O_2) carriers (HBOCs) restricted their application as red blood cell (RBC) substitutes. Among the myriad of strategies to produce Hb, using glutaraldehyde crosslinking to form polymerized Hb (PolyHb) remains popular due to its low cost and excellent scalability. However, the commercially developed PolyHbs, HBOC-201® (Biopure Corp, Cambridge, MA, USA) and PolyHeme® (Northfield Laboratories Inc., Northfield, IL, USA), primarily contained fractions of material at or below 250 kDa in MW. These low MW PolyHb and unpolymerized Hb are able to extravasate into the interstitial space where they scavenge nitric oxide (NO), resulting in vasoconstriction, systemic hypertension and oxidative tissue injury. Recent strategies to mitigate hypertension have used adenosine and nitroglycerine to attenuate the hypertensive response of HBOC-201®. While effective at controlling systemic hypertension, administration of adenosine and nitroglycerine must be carefully controlled to prevent hypotension and cannot address HBOC extravasation into the tissue space. Instead, we propose that employing the naturally occurring mechanism of Hb detoxification is a potential strategy to mitigate systemice hypertension resulting from administration of low MW PolyHb. The naturally occurring Hb scavenging protein, haptoglobin (Hp), plays a pivotal role in detoxifying stroma free Hb in the blood. Recently, we determined that Hp preferentially binds to low MW PolyHb (MW < 256 kDa). By binding to Hp, the molecular diameter of low MW PolyhHb is effectively increased. This increase in size may reduce tissue extravasation and subsequent NO scavenging of PolyHb.

Methodology: In this study, we synthesized a low MW PolyHb. The hydrodynamic diameter, O_2 affinity, and size distribution were measured. The Hp used for this study was purified from human Cohn Fraction IV derived from pooled human plasma. The resulting Hp contained both Hp2-1 and Hp2-2 phenotypes. Hp binding to PolyHb was confirmed *in vitro* with stop-flow fluorescence spectrometry and size exclusion chromatography. Hp was stabilized by binding it to heme-free Hb (apoHb) in the form of an apoHb-Hp complex. Systemic parameters, including heart rate (HR) and mean arterial pressure (MAP), were used as indicators of reduced hypertension. Additionally, intravital microscopy was used to examine how co-administration of apoHb-Hp influences functional capillary density (FCD), vascular tone, and blood flow.

Results and Conclusions: Polymerization of Hb resulted in decreased O_2 affinity and increased MW. PolyHb contained a significant fraction of low MW species (< 250 kDa). Hp was able to bind low MW PolyHb. Hp binding resulted in a significant increase in the average size of PolyHb. In the apoHp-Hb group, there was negligible change in MAP, FCD, HR, and microhemodynamics compared to baseline conditions. When compared to the systemic and microcirculatory changes observed in the apoHb and saline groups, the relatively small changes in the apoHb-Hp group indicate that Hp-based species may serve as materials to counteract the pressor effects of low MW HBOCs that are capable of binding to Hp. Thus, the administration of a Hb scavenging apoHb-Hp solution maintained physiologic hemodynamics during the transfusion of PolyHb.

Copyright © American Institute of Chemical Engineers. All rights reserved.

602372 Bacterial Extraction from Blood through Lysis Filtration and Subsequent Removal from Filter

Abstract:

Blood stream infections (BSIs) are challenging infections to rapidly diagnose. The current clinical diagnostic methods for BSIs require culturing the blood sample prior to identifying the bacteria and any antibiotic resistance the bacteria may contain. Removing the culturing step from the bacterial identification process of a BSI provides a significant reduction in the processing time, but shifts the difficulty from processing time to concentration, since clinical concentration levels can be as low as 10 CFU/mL in blood. This study developed a new method, whole blood filtering, to identify bacteria from a BSI without culturing the bacteria. Filtering whole blood achieved 100% bacterial removal from 5 mLs of whole blood in ~90 s. However, the bacteria were difficult to remove from the filter, leading to further investigation. At a very low bacterial concentration of 200 CFU/mL, a blood lysis solution of 3% Tween 80 followed by a 3% Pluronic F108 backflush solution achieved 60% removal of the bacteria from the filter.

Copyright © American Institute of Chemical Engineers. All rights reserved.

603159 *Project SAVE: Silica Adsorption Vaccine Encapsulation*

Abstract:

It is estimated that 50% of vaccines produced annually are wasted because effectivity is dependent on protein structure and heat exposure disrupts the intermolecular interactions that maintain this structure. Since 90% of vaccines require a temperature-controlled supply chain, it is necessary to create a cold chain system to minimize vaccine waste. Influenza viruses cause seasonal flu epidemics that affect millions worldwide, and cause especially serious illness to children, elderly and those with compromised immune systems. Globally, influenza causes 390,000 to 650,000 deaths each year. We are developing a more sustainable technology via the adsorption of Quadrivalent Influenza Vaccine (QIV) onto mesoporous silica gels, improving the thermal stability of protein-based therapeutics. A solution depletion method in combination with UV-Vis is used to study the adsorption of QIV onto silica gels. The silica-QIV complex is heated above the denaturing temperature of the protein and then the QIV is removed using a detergent. Pore diameter, pore volume and surface area were characterized for different silica gels. In general, silicas with larger pore volumes and pore diameters tend to adsorb more QIV. Silica adsorption isotherms were created in order to calculate the maximum QIV monolayer coverage and the kinetic rates of adsorption and desorption. Circular Dichroism (CD) analyses confirmed that the adsorbed QIV after the heat treatment displayed a similar "W" shape CD signal as the native QIV, indicating the conservation of α-helices. In contrast, the unprotected QIV after being exposed to high temperature shows a flat CD signal, demonstrating the loss of secondary structure. In conclusion, we have successfully increased the thermo-tolerance for QIV using mesoporous silica and continue to further optimize mesoporous silica's physiochemical properties to improve adsorption and desorption yields.

Copyright © American Institute of Chemical Engineers. All rights reserved.

606784 Exploring Hemolysis Due to Suture Dehiscence in Prosthetic Heart Valves

Abstract:

Hemolysis is a serious medical condition in which red blood cells (RBCs) lose hemoglobin and are removed from the circulation. One known cause of this condition is exposure of RBCs to high shear stresses, including those created by intravascular prosthetics such as artificial heart valves. In 1975, as many as 15% of patients with artificial heart valves had severe hemolytic anemia[1], but more modern designs and improvements to surgical technique have brought that number below 1%. The anemia caused by modern devices is often attributable not to the device itself, but rather to paravalvular leakage. When this leakage occurs, a small opening appears between the normal heart tissue and the rim of the prosthetic, creating a channel through which blood can flow as a jet in the improper direction from the lower chamber to the upper chamber of the heart. This condition, called dehiscence, which typically occurs when a suture stitching pulls out of the tissue, subjects RBCs to shear stresses outside the normal physiological range, and can lead to an increased rate of hemolysis and hemolytic anemia.

The relationship between flow parameters and the hemolysis index (HI), a measurement of the degree of hemolysis occurring in a blood sample, has been the object of some study. Leverett et al.[2] made an observation in 1972 that has allowed for empirical correlations between the shear stress, the exposure time, and the HI for laminar flows. For the turbulent flow regime, prior work done by Ozturk et al.[3] in this research group matched the HI to the Kolmogorov Length Scale (KLS) size of turbulent eddies.

An anatomically representative model of a human left atrium was constructed in part using a software package developed by Xiong et al.[4] which transcribed MRI slices into epicardium outlines. These outlines were then stacked together using Free-D[5], a specialized software program designed to connect several cross sections into a 3D object. The resulting file was imported to ANSYS Fluent 19.1[6] for use in computational fluid dynamics (CFD) calculations. A 3 mm long, semi-circular channel of diameter 1 mm was added to the atrium to simulate the paravalvular leakage. A Casson constitutive equation was used to describe variable blood viscosity.

Ozturk[3] describes a process known as eddy analysis that is used to connect turbulence to the HI. The KLS, defined in this case as the fourth root of the ratio of the kinematic viscosity of blood to the turbulent kinetic energy dissipation rate, is calculated in the volume of the atrium including the dehiscence site. The total volume occupied by fluid structures that correspond to KLS eddies comparable to the size of RBCs is determined. Eddies up to a KLS of 10μm, roughly the same diameter as RBCs, are shown to have significant impact on hemolysis, since eddies of such dimensions apply greater shear to RBCs and have been shown to be a main cause of RBC trauma.

This eddy analysis was preformed on the atrium model to predict the hemolysis that occurs under paravalvular leakage conditions. The goal is to be able to evaluate how severe the problem is when it occurs in a patient, potentially preventing an unnecessary and life-threatening post-op surgery, or recommending a necessary and life-saving repair.

References:

1. Sethi, P.; Murtaza, G.; Rhaman, Z.; Zaidi, S.; Helton, T.; Paul, T. "Valvular Hemolysis Masquerading as Prosthetic Valve Stenosis," Cureus. 9, e1143 (2017).
2. Leverett, L. B., Hellums, J. D., Alfrey, C. P., and Lynch, E. C., "Red Blood Cell Damage by Shear Stress," Biophys. J., 12(3), pp. 257–273 (1972).
3. Ozturk, M., Papavassiliou, D. V., O'Rear, E. A., "An approach of Assessing Turbulent Flow Damage to Blood in Medical Devices," J. Biomech Eng. 139, 011001-011008 (2017).
4. Xiong, Z., Fedorov, V. V., Fu, X., Cheng, E., Macleod, R., & Zhao, J. "Fully Automatic Left Atrium Segmentation From Late Gadolinium Enhanced Magnetic Resonance Imaging Using a Dual Fully Convolutional Neural Network," *IEEE transactions on medical imaging, 38*(2), 515-524. (2019).
5. Andrey P & Maurin Y "Free-D: an integrated environment for three-dimensional reconstruction from serial sections". Journal of Neuroscience Methods, 145, 233-244 (2005).
6. Ansys® Fluent, Release 19.1.

Copyright © American Institute of Chemical Engineers. All rights reserved.

609676 Lobe-Specific Aerosol Targeting in a 3D Printed Lung Model

Abstract:

Adverse pathology in respiratory diseases such as cancer or pneumonia are localized to regions of the lung, yet current inhalation devices fail to provide controlled regional specificity of delivered therapeutics. Computational fluid dynamics (CFD) simulations have shown feasibility of directing deposition to a lung, lobe, or region,[1] but there have been limited experimental assessments of this targeting approach. Our objective is to develop a lung model and inhalation device to harness local streamlines and direct the aerosol to a lobe of interest, resulting in improved deposition profiles, increased drug concentrations, and minimized off-target effects. We fabricated hollow, patient-derived lung mimics developed from a computed tomography (CT) scan and 3D printed using a Carbon M1 printer and simulated breathing using a vacuum pump. Guided by CFD experiments, a custom "regional targeting device" (RTD) was created to control aerosol inlet location. Sampling ports were placed at the outlet of each lobe to assess distribution of monodisperse polystyrene particles aerosolized in a Collison jet nebulizer at varied flow rates. Airflow profiles that mimic naturally occurring asymmetric inspiratory lung airflow distributions were successfully recreated in our model, in agreement with analogous CFD simulations and *in vivo* data.[2] Inlet locations were identified for each lobe that skewed deposition toward the targeted lobe and away from non-targeted lobes. For example, when targeting the lower left lobe, deposition was increased from 23.7% before targeting to 86% with the RTD, a 4x increase with a p-value of 0.002. Corresponding deposition to other lobes decreased significantly, representing decreased off-targeting. Increased targeting efficiency was observed with smaller particles and decreased flow rates. Increasing particle size and flow rate spreads deposition to nearby lobes. This approach can result in dramatic concentration of inhaled therapeutics to a region of interest, future work will continue to evaluate the success of this approach across multiple patient geometries of varied age.

References:

1. Feng, Yu, et al. "An *in silico* subject-variability study of upper airway morphological influence on the airflow regime in a tracheobronchial tree." *Bioengineering*4.4 (2017): 90.

2. Sul, Bora, et al. "Assessing airflow sensitivity to healthy and diseased lung conditions in a computational fluid dynamics model validated in vitro." *J Biomech Eng-T* 140.5 (2018): 051009.

Copyright © American Institute of Chemical Engineers. All rights reserved.

610887 Concurrent Mechanical Stretching and Co-Cultivation for Promoting the Chondrogenesis of Human Adipose-Derived Stem Cells

Abstract:

Bones in articulating joints are cushioned with a viscoelastic tissue called Articular Cartilage (AC). AC acquires a unique mechanical structure that absorbs bodyweight loads on the knee vascularization. However, because AC is avascular in nature, cartilage lacks the intrinsic ability to regenerate but rather degenerates with wear and tear, causing osteoarthritis (OA). OA symptoms include pain, swelling, loss of flexibility, and, eventually, disability. Currently available treatments, including surgeries like total knee arthroscopy, alleviate the pain, but they can never be used to restore full joint mobility. Patients with OA never return to a completely healthy lifestyle even after treatment, and they become prone to developing severe additional health problems due to the inactivity. Articular cartilage tissue engineering (ACTE) is an emerging field in which regenerative options for AC are being developed. Although the field has potential, ACTE is critically limited by the availability of chondrogenic cell sources. In this work, the influence of combined co-cultivation, a biochemical stimuli, and cyclic tensile stretching (CTS), a mechanical stimuli, on chondrogenesis is investigated. The viability and biochemical content of human adipose-derived stem cells (hASCs) co-cultivated with human chondrocytes (hAChs) under 10% CTS at 1Hz for 3 h/day for 14 days are compared to statically cultured hASCs. Our findings indicate that collagen secretion was enhanced by at least 1.7-fold with co-cultivation, 1.8-fold with CTS, and by 6.2-fold with co-cultivation under CTS, when compared to static cultures of hASCs.

Copyright © American Institute of Chemical Engineers. All rights reserved.

Hybrid Electro-Plasmonic Neural Stimulation and Its Implications for Prosthetic Devices

Ratka Damnjanovic, Parveen Bazard, Venkat Bhethanabotla and Robert Frisina, University of South Florida, Tampa, FL

Abstract Text:

Purpose/Introduction:

Electrical stimulation tends to spread to surrounding tissues, making it difficult to stimulate discrete neural sites with traditional neural prosthetic devices, such as cochlear implants, cardiac pacemakers, electromyography stimulators, sciatic nerve stimulators, etc. used for neural stimulation. To facilitate specific point stimulation, various nanomaterial-assisted neural stimulation approaches have been reported in recent years[1-12] where different localized fields are activated (electric, magnetic, thermal) employing different nanomaterials for stimulation. Infrared stimulation has been shown to excite neurons, but with limitation of heating surrounding tissue[13-14]. In a previous study[15] we demonstrated the use of visible light (532 nm) laser pulses to trigger a localized surface plasmon resonance (LSPR) effect from the gold nanoparticles coated microelectrodes for stimulating plasmonic modulation of SH-SY5Y neuroblastoma cells response, but with a limited success rate and detrimental effects on the cell membrane with higher levels of pure optical stimulation. To address the limitations, we developed a novel platform for spatially and temporally precise neural excitation, using hybrid electro-plasmonic stimulation technique in a whole-cell patch-clamp configuration to elicit electrical responses in primary trigeminal neurons.

Experimental Design/Methods:

Borosilicate glass micropipettes were silanized with 10 % solution of γ-aminopropyl triethoxy silane in ethanol, following the Turkevich method[16]. Then, the tip of the micropipette was dip-coated with 20 nm-diameter colloidal gold nanoparticles (Au NPs) for making the gold-coated micropipettes[15]. These micropipettes were used for producing the needed plasmonic stimulation effect for the *in-vitro* neuron stimulations. 5-7 weeks C57B1/6 mice were used for the study. Mouse was decapitated, and the trigeminal ganglia nerve was removed. The nerve tissue was dissociated enzymatically, with HBSS containing collagenase type I and dispase II, and finally neurons were cultured in L15 media containing 10 % FBS and left to incubate in an incubator maintained at 37 C, 5 % CO_2 over night after dissociation. All cells were used within 36 hours timeframe. All the electrophysiology experiments were done using whole-cell patch-clamp methods. Microelectrode was placed next to a single patched neuron cell and a 532 nm green laser beam was focused onto the gold coated electrode tip using an optical fiber. Whole cell patch-clamp techniques were used in conjunction with an Axopatch 700B amplifier, Digidata 1440 interface, and pCLAMP-9 software (Axon Instruments, Union City, CA, USA). Extracellular solution was used to flood the cells in the petri dish. For the hybrid stimulation, electrical stimulus was added, in addition to the optical plasmonic stimulation. Neural responses were recorded using the patch electrode.

Major Findings/Results:

Electrical stimulation was used before and after the optical stimulation to verify that cells are healthy and can fire action potentials. Initially, experiments were done with pure optical stimulations (1-5 ms pulse, 100-120 mW power). Plasmonic stimulations were more detrimental to the cells as compared to pure electrical stimulation, with success rate of firing action potential less than 30 % of stimulated cells. Also, pure optical stimulations led to detrimental effects on the cell membrane. To overcome these issues, we combined electrical and optical stimulation to add the advantages from both stimulation methods, plasmonic and electrical, while decreasing the disadvantages from both methods when used alone. The percentage reduction in input current with hybrid stimulation, as compared to the current required to fire action potential with pure electrical stimulation, was approximately 40%. Also, electrical action potentials were recorded at a higher success rate with hybrid stimulation as compared to pure plasmonic stimulation (83% vs. 26%). Neuron cells survival and viability after hybrid stimulation was superior to that of pure optical stimulation (72% vs. 13%). We were also able to achieve multiple consecutive action potentials firings when applying consecutive hybrid stimulation on the cells. Further optimization of the lead and lag time of the electrical vs. plasmonic stimulus was also studied. The most optimal output is achieved when electrical stimulus leads before optical by 0.4 ms and not more than 1.4 ms, as well as when optical stimulus leads electrical by 0.6ms or less. The best repeatability of the action potential firing was achieved when electrical leads before optical by less than 1 ms.

Summary:

We demonstrated that with our hybrid stimulation platform, a reduction of up to ~40% of the input current threshold is achieved when optical stimulation is added to the sub-threshold electrical stimuli. Nanomaterials, specifically gold, maximize the utility of thermal stimulation via the LSPR phenomena in the triggering action potentials. Our findings show the applicability of short duration pulses (1-5 ms) when applied repeatedly, for sub-threshold electrical and LSPR visible light stimulation pulses, in combination with AuNPs-coated substrates (nanoelectrodes), can reliably trigger a train of action potentials, for obtaining repeatable multiple trains of APs from neurons. Neural cell survival rates and viability after hybrid stimulation was superior to that of pure optical stimulation. The input current sufficient to trigger APs with multiple hybrid stimulation was 35-40% lower, matching what was observed earlier with single AP recordings. This reinforces the previous findings above, supporting the effectiveness of the proposed platform for hybrid electro-plasmonic stimulation of neurons, which opens doors of opportunities to develop a new generation of high-acuity neural modulation prosthetic devices, tunable for the individual patient's needs. Our long-term goal is to develop nanoparticle-light based cochlear implants for the hearing-impaired population. The reduction of current required to trigger action potentials, and the evidence that cells stay healthy after repeated exposure to hybrid stimulation are ground breaking results for a new generation tunable cochlear implants that can offer more precise frequency modulation enabled by more selective activation of auditory neurons.

Reference:

[1] Wang, Y.; Guo, L., Nanomaterial-Enabled Neural Stimulation. *Front Neurosci* **2016,** *10*, 69.

[2] Deisseroth, K., Optogenetics: 10 Years of Microbial Opsins in Neuroscience. *Nat Neurosci* **2015,** *18* (9), 1213-1225.

[3] Colombo, E.; Feyen, P.; Antognazza, M. R.; Lanzani, G.; Benfenati, F., Nanoparticles: A Challenging Vehicle for Neural Stimulation. *Front Neurosci* **2016,** *10*, 105.

[4] Carvalho-de-Souza, J. L.; Treger, J. S.; Dang, B.; Kent, S. B.; Pepperberg, D. R.; Bezanilla, F., Photosensitivity of Neurons Enabled by Cell-Targeted Gold Nanoparticles. *Neuron* **2015,** *86* (1), 207-17.

[5] Marino, A.; Arai, S.; Hou, Y.; Sinibaldi, E.; Pellegrino, M.; Chang, Y. T.; Mazzolai, B.; Mattoli, V.; Suzuki, M.; Ciofani, G., Piezoelectric Nanoparticle-Assisted Wireless Neuronal Stimulation. *Acs Nano* **2015,** *9* (7), 7678-89.

[6] Chen, R.; Romero, G.; Christiansen, M. G.; Mohr, A.; Anikeeva, P., Wireless Magnetothermal Deep Brain Stimulation. *Science* **2015,** *347* (6229), 1477-80.

[7] Eom, K.; Kim, J.; Choi, J. M.; Kang, T.; Chang, J. W.; Byun, K. M.; Jun, S. B.; Kim, S. J., Enhanced Infrared Neural Stimulation Using Localized Surface Plasmon Resonance of Gold Nanorods. *Small* **2014,** *10* (19), 3853-3857.

[8] Yong, J.; Needham, K.; Brown, W. G.; Nayagam, B. A.; McArthur, S. L.; Yu, A.; Stoddart, P. R., Gold-Nanorod-Assisted Near-Infrared Stimulation of Primary Auditory Neurons. *Adv Healthc Mater* **2014,** *3* (11), 1862-8.

[9] Chen, S.; Weitemier, A. Z.; Zeng, X.; He, L.; Wang, X.; Tao, Y.; Huang, A. J. Y.; Hashimotodani, Y.; Kano, M.; Iwasaki, H.; Parajuli, L. K.; Okabe, S.; Teh, D. B. L.; All, A. H.; Tsutsui-Kimura, I.; Tanaka, K. F.; Liu, X.; McHugh, T. J., Near-Infrared Deep Brain Stimulation *via* Upconversion Nanoparticle-Mediated Optogenetics. *Science* **2018,** *359* (6376), 679-684.

[10] Yoo, S.; Hong, S.; Choi, Y.; Park, J. H.; Nam, Y., Photothermal Inhibition of Neural Activity With Near-Infrared-Sensitive Nanotransducers. *Acs Nano* **2014,** *8* (8), 8040-9.

[11] Li, W.; Luo, R.; Lin, X.; Jadhav, A. D.; Zhang, Z.; Yan, L.; Chan, C. Y.; Chen, X.; He, J.; Chen, C. H.; Shi, P., Remote Modulation of Neural Activities *via* Near-Infrared Triggered Release of Biomolecules. *Biomaterials* **2015,** *65*, 76-85.

[12] Pappas, T. C.; Wickramanyake, W. M.; Jan, E.; Motamedi, M.; Brodwick, M.; Kotov, N. A., Nanoscale engineering of a cellular interface with semiconductor nanoparticle films for photoelectric stimulation of neurons. *Nano Lett* **2007,** *7* (2), 513-9.

[13] Shapiro, M. G.; Homma, K.; Villarreal, S.; Richter, C.-P.; Bezanilla, F., Infrared Light Excites Cells By Changing Their Electrical Capacitance. *Nature Communications* **2012,** *3*, 736.

[14] Wells, J.; Kao, C.; Mariappan, K.; Albea, J.; Jansen, E. D.; Konrad, P.; Mahadevan-Jansen, A., Optical stimulation of neural tissue in vivo. *Optics letters* **2005,** *30* (5), 504-506.

[15] Bazard, P.; Frisina, R. D.; Walton, J. P.; Bhethanabotla, V. R., Nanoparticle-based Plasmonic Transduction for Modulation of Electrically Excitable Cells. *Scientific Reports* **2017,** *7* (1), 7803.

[16] Turkevich, J.; Stevenson, P. C.; Hillier, J., A study of the nucleation and growth processes in the synthesis of colloidal gold. *Discussions of the Faraday Society* **1951,** *11*, 55-75.

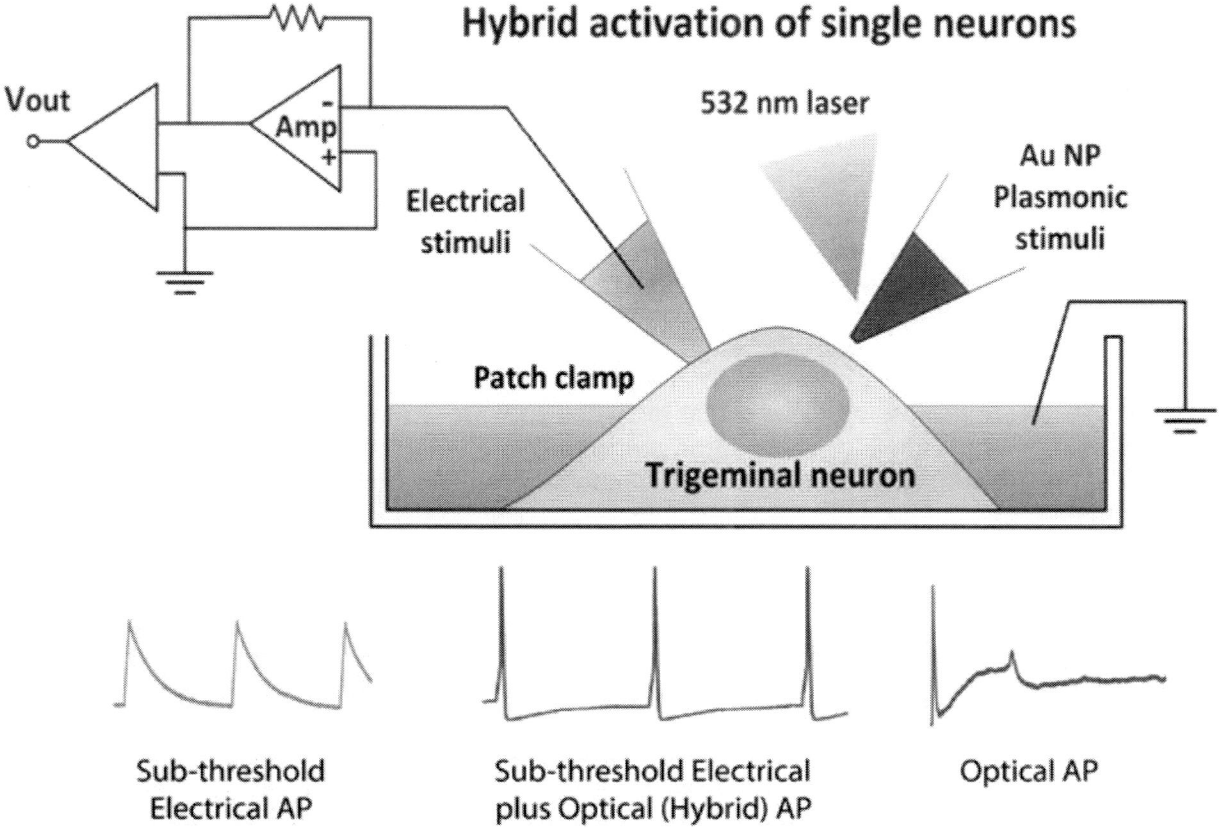

609353 Artificial Intelligence to Improve Blood Glucose Control for People with Type 1 Diabetes

Abstract:

An individual with type 1 diabetes (T1D) must be vigilant in monitoring their blood glucose values while considering the impact of meals, exercise, and other critical events. More and more people with T1D are wearing continuous glucose monitors (CGM, providing blood glucose values at 5 min intervals) and insulin pumps that continuously deliver rapid-acting insulin. However, only a few commercial closed-loop systems, that automatically adjust insulin in response to glucose changes, are currently available. These systems require manual meal announcements to provide insulin boluses at mealtime (feedforward control) and automatically suspend insulin when glucose is low. Still, missed meal announcements and exercise-induced hypoglycemia are major causes for suboptimal glucose control in people with T1D. Another common problem is CGM pressure-induced sensor attenuation (PISAs) that typically occur overnight when a user sleeps on their CGM, resulting in inaccurate CGM signals. PISAs can cause the pump to shut off unnecessarily. Evaluating the likelihood of CGM readings being PISAs can allow a controller to ignore readings with a high PISA probability.

Automatic activity detections and forecasts can allow a controller to prompt for meals and anticipate exercise. To this end, we have developed an automatic activity detection system using a smartwatch and smartphone. Long Short-Term Memory (LSTM) networks, convolutional neural networks (CNN), probabilistic graphical models (PGM), and lookup tables are used to create minute-based probabilities of other, sleep, eating, exercise, and null using sensor data. A particle filter is used to detect activity events by aggregating minute-based probabilities. Population level time of day prior probabilities are used to forecast exercise and sleep. Support vector machines (SVM), random forests, and neural networks are compared for classification of PISAs.

The CNN achieves a class-weighted accuracy of 85% using smartwatch sensor readings. The particle filter detects 99.9% of sleep, 99.4% of eating, and 94.4% of exercise events. 4.3 false sleep events/day, 2.5 false eating events/day, and 0.38 false exercise events/day are detected. Time of day probabilities provide higher likelihoods of impending exercise and sleep when they truly occur compared to when they do not. For PISA classification, the random forest achieves a true positive rate of 90% and a false positive rate of 6.3%.

Reliable activity detections, and forecasts of exercise and sleep can be achieved with daily patient review. Using a random forest, reliable likelihoods of PISAs can be obtained. Signals from the CGM and insulin pump can therefore be supplemented with activity detections/forecasts and PISA detections to improve glucose control. Furthermore, these techniques can reduce patient burden and allow people with T1D to lead more normal lives.

Copyright © American Institute of Chemical Engineers. All rights reserved.

598365 Optical and Electronic Monitoring of Ion Channels Implicated in Pain Sensation Using Biomembrane Bioelectronic Devices

Abstract:

Protein ion channels in cell membranes are a key elements in pain signaling. Around 80 ion channels have strong links to pain sensation, but only a small number of these channels have been successfully drugged for pain management. The opioid crisis only exacerbates the need to understand the underlying causes of pain and find new targets to manage it with minimal side effects. One reason for the bottleneck in discovery of new drugs is the inadequacy of conventional platforms for drug discovery that monitor modulation in ion channel activity. Current methods require a skilled hand, yet are still very laborious, time-consuming, and low throughput. There is an urgent need to develop faster, authentic, and high throughput assays that give direct and quantitative information on ion channel function, drug interactions, and insight on the efficacy of a developed therapy. This presentation will highlight a new technology for the rapid and sensitive measurement of subtle changes in cell membrane permeability and capacitance in response to small molecule interactions with protein ion channels embedded in it. This device features a biomimetic membrane that provides an authentic lipid environment, native cell membrane material to support the natural transmembrane ion channel function, and a sensor made from an electroactive polymer surface upon which the biomembrane is supported. Of focus here is the ATP-gated P2X family ion channels and Trek ion channels. Both ion channel receptors are considered important therapeutic targets for multiple pain modalities including neuropathic, inflammatory, and chronic pain states. This biomembrane sensor is capable of reading and transducing ion fluxes through these ion channels as it depends on the availability of ATP, the presence of specific biomolecules, and drug compounds. Critically, this new sensing platform captures these essential features: 1) it preserves native plasma membrane, protein orientation, fluidity, and activity; 2) is robust and long-lasting; and 3) is compatible with high throughput expansion. The effective treatment of pain remains a challenge as current therapeutics still lack desired levels of efficacy and tolerability. Given the considerable amount of research focus and development on identifying drug targets and engineering new drugs for the treatment of pain, this new biosensing platform may prove to be a useful new approach for drug screening.

Copyright © American Institute of Chemical Engineers. All rights reserved.

610704 Tetracosane Functionalized Nanostructure Titanium Oxide Sensors for Screening Pneumonia from Breath

Abstract:

Volatile organic compounds (VOCs) are present in the environment and human bodies. Ethanol, acetone, and isoprene are examples of VOCs in the healthy human body, specifically in breath[1]. Diseases such as pneumonia have VOCs associated with it in breath[2]. Pneumonia is the cause of 16% death of children under five globaly[3]. In the US, pneumonia is the first common reason of hospitalization in children and the second common reason for adults' admission after women giving birth[3]. Therefore, there is a need for an inexpensive, noninvasive, and quick pneumonia screening method that can simply be performed in clinics on the first patient visit. Patients with pneumonia have been shown to have breath profiles that include unique VOCs that derive from the biological pathogen and perturbations in the health of a patient. The presence of heptane and increase in the amount of ethanol are examples of VOCs associated with pneumonia[2]. Developing a device to measure the presence of these VOCs in breath could be used as a screening method to detect pneumonia. In this work, a suitable molecule is functionalized on a titanium oxide platform to give an electrochemical response in the presence of pneumonia VOCs.

The sensor synthesis process consists of anodization of titanium oxide nanotubes and annealing the anodized self-ordered nanotubes under oxygen. Titanium oxide nanotubes provide an adequate conductivity and have an amorphous surface area suitable for external coatings. As mentioned earlier, ethanol is at a higher concentration in the breath of patient's suffering from pneumonia compared healthy human's breath[2]. Titanium oxide nanostructure at given voltages responds to the presence of ethanol by showing current changes. But for the purpose of detecting pneumonia, a more specific sensor is needed. To design a sensor that also responds to heptane, titanium oxide nanotubes were functionalized with tetracosane. Tan et al.[4] showed that when alkane gets mixed with tetracosane on a conductive surface. At a constant voltage of 1.5V, the tetracosane functionalized titanium oxide sensor performs as a chemiresistive sensor. It responds to heptane and ethanol in the gas phase by showing current changes. Table (1) shows the current response for a tetracosane functionalized titanium oxide sensor to ethanol, heptane and humid air exposure. Due to the presence of humidity in the human breath, the sensor was exposed to controlled humid samples. When heptane, humid air, and ethanol were exposed to the functionalized sensor, the current increased to different magnitudes (figure 1). Room air was bubbled through a solution of the VOC of interest and exposed to the sensor that was placed on PCB board. The current change magnitude and the response time of the sensor varied for each exposure. The sensor's response to humid air and ethanol was instant, whereas the coated tetracosane sensor responded to heptane exposure after a minimum of 40 seconds. The sensor's current change to ethanol exposure was a minimum of three orders of magnitude higher than heptane in the first 100 seconds of sample exposure. The tetracosane functionalized sensor shows a different response to each heptane, ethanol, and humid air in gas phase. The sensor's response to different concentrations of heptane and ethanol separately and mixed together in mimicked breath samples will be presented.

References:

[1] J. D. Fenske and S. E. Paulson, "Human breath emissions of VOCs," *J. Air Waste Manag. Assoc.*, vol. 49, no. 5, pp. 594–598, 1999.

[2] R. Schnabel et al., “Analysis of volatile organic compounds in exhaled breath to diagnose ventilator-associated pneumonia,” Sci. Rep., vol. 5, no. July, pp. 1–10, 2015.

[3] N.A., “Top 20 Pneumonia Facts 2018,” Am. Thorac. Soc., p. 2018.

[4] Jiunn-Liang Tan; Zheng-Xin Yong; Chong-Kin Liam, Using a chemiresistor-based alkane sensor to distinguish exhaled breaths of lung cancer patients from subjects with no lung cancer. *Journal of Thoracic Disease* 2016, 10.2103

	Heptane	Water	Heptane	Ethanol
Current Change (uA)	2.19e-2	3.04e-2	4.53e-2	7.41e+1

Table (1), Sensor's current changes when heptane, water, and ethanol was exposed to it, at 1.5V.

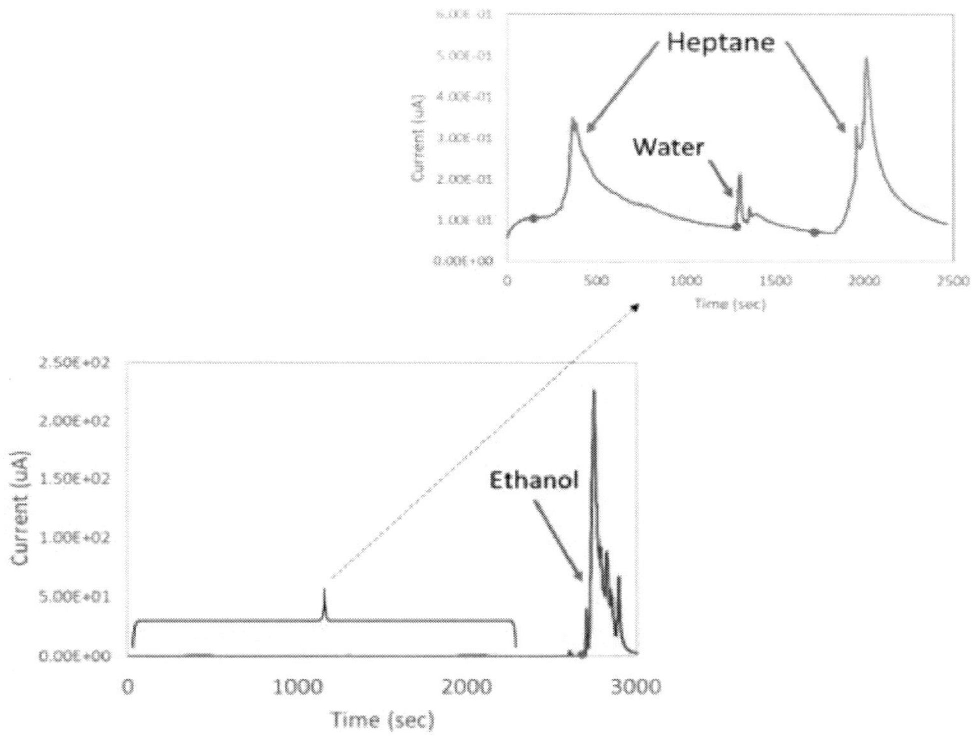

Figure (1): An example of tetracosane sensor response to Heptane, water and ethanol. The Blue dots are the starting time the sensor was exposed to the sample.

Copyright © American Institute of Chemical Engineers. All rights reserved.

603401 Fabrication of Enzymatic Electrochemical Biosensors for Real-Time Nicotine Detection

Abstract:

Biosensors provide an output signal based upon a successful recognition of an analyte. Key to this process in oxidase-based biosensor is analyte recognition by an enzyme and subsequent signal transduction. As such, the sensor design process must successfully address a series of requirements associated with the interface between the enzyme and an electrode. Here we present the development, assessment, and optimization of a nicotine sensor based on a nicotine catabolizing redox enzyme and Prussian Blue coated screen-printed electrodes to afford a chronoamperometric response proportional to nicotine. An optimized polymeric matrix immobilization method maintains enzymatic activity on the mediator-coated screen-printed electrode and enables quantitative detection of nicotine. The resulting biosensor detects nicotine in urine with high sensitivity with a limit of detection of 34 μM over the range of 0-200 μM.

Copyright © American Institute of Chemical Engineers. All rights reserved.

611271 Point-of-Care Fungal Pathogen Identification Using Surface Acoustic Waves Coupled with Metal-Enhanced Fluorescence

Abstract:

Corneal ulcers, typically caused by infection, are among the leading causes of vision loss and require prompt intervention. Without effective treatment, permanent vision impairment[i] and more serious infections[ii] can result. Rather than spending days culturing the pathogen to determine whether the infection is bacterial or fungal, antibiotics are typically applied as an empiric treatment[iii]; as the incidence of fungal infections increases, however, complications are becoming more prevalent[iv]. The two types of infection may instead be rapidly distinguished at point-of-care by means of a self-contained acoustic isolation and detection system, allowing doctors to immediately prescribe the appropriate remedy. The core of this system is a piezoelectric chip made from a material such as quartz or $LiNbO_3$ whose surface is functionalized with a chitin-binding lectin having a strong affinity for the cell walls of fungal pathogens. By electrically exciting this chip, Rayleigh waves lyse pathogen cells, mix the sample, and remove nonspecifically-bound materials[v], purifying chitin and isolating it on the chip's surface. The isolated analyte is then detected using a chitin-sensitive fluorescent probe. The detection limit may be further improved by means of metal-enhanced fluorescence using a thin layer of silver nanostructures formed by rapid thermal annealing that couple with the probe signal[vi]. This technique has proven effective with other analytes at concentrations on the order of picograms per milliliter[vii].

[i] McClintic, S. M., et al. "Improvement in corneal scarring following bacterial keratitis." *Eye* 27.3 (2013): 443-446.

[ii] Scott, Ingrid U., et al. "Endophthalmitis associated with microbial keratitis." *Ophthalmology* 103.11 (1996): 1864-1870.

[iii] Collier, Sarah A., et al. "Estimated burden of keratitis—United States, 2010." *MMWR. Morbidity and mortality weekly report* 63.45 (2014): 1027.

[iv] Tuberville, Audrey W., and Thomas O. Wood. "Corneal ulcers in corneal transplants." *Current eye research* 1.8 (1981): 479-485.

[v] Cular, Stefan, et al. "Removal of nonspecifically bound proteins on microarrays using surface acoustic waves." *IEEE Sensors journal* 8.3 (2008): 314-320.

[vi] Morrill, Samuel, Venkat Bhethanabotla, and Mandek Richardson. "Biomarker quantification at clinically relevant concentrations using metal enhanced fluorescence combined with surface acoustic waves." *SENSORS, 2014 IEEE*. IEEE, 2014.

[vii] Liu, Jun, Shuangming Li, and Venkat R. Bhethanabotla. "Integrating metal-enhanced fluorescence and surface acoustic waves for sensitive and rapid quantification of cancer biomarkers from real matrices." *ACS sensors* 3.1 (2018): 222-229.

Copyright © American Institute of Chemical Engineers. All rights reserved.

606858 A Power-Free Colorimetric Biosensor for Detection of Mycobacterium Tuberculosis

Abstract:

Tuberculosis (TB) remains a significant health threat in the human society. Currently, about one-third of the human population is ill with TB worldwide due to the lack of appropriate technologies for treatment and prevention [1]. *Mycobacterium tuberculosis* is the etiological agent of tuberculosis. TB bacteria enter the body by inhalation of transmittable droplet nuclei. There are different clinical manifestations of tuberculosis such as weight loss, low-grade fever, and drenching night sweats [1].

Early diagnosis plays a major role in controlling the transmission of TB. Traditional microbial culture-based tests are the most common procedures currently used which involve cell culture, cell counts, and cell enrichment, however, this method is time consuming and difficult, particularly for the slow-growing bacteria like *M. tuberculosis*. The global public health problem of TB worldwide demands for the development of more rapid and sensitive detection methods. Currently, many methods and techniques have been developed for rapid detection of *M. tuberculosis*, such as polymerase chain reaction (PCR) [3], enzyme-linked immunosorbent assay (ELISA) [4], flow cytometry and the latest WHO-endorsed GeneXpert based test [5]. Nevertheless, they cannot provide the detection results in real-time and most of these methods are often difficult and complex to execute and centralized in large stationary laboratories because the difficulties of the operation would typically require professionals that are often expensive to retain [6]. As a result, the development of portable, real-time, sensitive, rapid, and accurate methods for *M. tuberculosis* detection is essential to effectively prevent TB infection.

Accordingly, the present study proposes a modified microchannel, a DNA nano-sensor and a portable detection system for rapid detection of mycobacterial tuberculosis enzyme activity. In the proposed detection method, a porous hydrogel was fabricated inside the microchannel via sol-gel chemistry. Polystyrene (PS) particles (45 μm) were self-assembled for 12 h under 30°C to form a closely-packed colloidal template. The poly(ethylene glycol) diacrylate solution was added dropwise on the PS template so that the solution filled the interstitial spaces within the self-assembled PS. The resulting hydrogel was cured for 30 min and then soaked in toluene for 24 h to form a porous hydrogel by removing the PS particles.

The DNA nano-sensor was prepared by conjugating a designed single DNA strand into carrier particles and AuNPs. The sensing mechanism was on the basis of cleaving hydrogen bonds between DNA strands with active MTopI and ligating the carrier particles and AuNPs. Samples containing active MTopI and DNA nano-sensor could flow along the modified microchannel driven by capillary force, at the middle of the microchannel a porous hydrogel with pore sizes smaller than the carrier particle diameter was embedded. When the MTopI was inactive, the carrier particles and AuNPs were not ligated. As a result, AuNPs were washed away from the microchannels. On the other hand, an active MTopI was capable of linking the carrier particles with AuNPs tended to aggregate in the front end of the porous hydrogel, resulting in a visible red color band in the microchannel. Readouts were performed using a smartphone-MATLAB algorithm interface. Overall, the proposed integrated platform presented in this study promotes point-of-care-testing and provides a compact and reliable tool for early diagnosis for TB patients in resource-limited areas.

- **REFERENCE**

1. WHO Global Tuberculosis Report, 2018.
2. E. Torres-Chavolla and E. C. Alocilja, "Aptasensors for detection of microbial and viral pathogens," Biosensors and Bioelectronics, vol. 24, no. 11, pp. 3175–3182, 2009.
3. D. Rodríguez-Lázaro, M. D'Agostino, A. Herrewegh, M. Pla, N. Cook, and J. Ikonomopoulos, "Real-time PCR-based methods for detection of Mycobacterium avium subsp. paratuberculosis in water and milk," International Journal of Food Microbiology, vol. 101, no. 1, pp. 93–104, 2005.
4. B. Patterson, C. Morrow, V. Singh, A. Moosa, M. Gqada, J. Woodward, V. Mizrahi, W. Bryden, C. Call, S. Patel, D. Warner, R. Wood Detection of Mycobacterium tuberculosis bacilli in bio-aerosols from untreated TB patients Gates Open Res, vol.1, pp. 11, 2017.
5. CM. Denkinger, SV. Kik, DM. Cirillo, M. Casenghi, T. Shinnick, K. Weyer, C. Gilpin, CC. Boehme, M. Schito, M. Kimerling, et al: Defining the needs for next generation assays for tuberculosis. J Infect Dis, vol. 211 Suppl 2, pp. S29-38, 2015.
6. W. ShuQi, I1. Fatih, D.L Gennaro, S. Amit, D. Utkan, Point-of-Care assays for Tuberculosis: Role of Nanotechnology/ Microfluidics. Biotechnol Adv., vol. 31(4), pp. 438–449, 2013.

Copyright © American Institute of Chemical Engineers. All rights reserved.

612133 Evaluations of Three Low-Cost PM Monitors Against a Reference PM Instrument

Abstract:

In recent years, low-cost particulate matter (PM) monitors have significantly drawn scientific attention for their affordable cost and feasibility. These low-cost monitors are used to measure PM number and mass concentrations in various size ranges and compositions. One of the main objectives of the current study was to evaluate the accuracy of three different low-cost monitors (PupleAir, Dylos, Speck) against a reference instrument (Scanning Mobility Particle Sizer (SMPS)) in the presence of particles from grilling beef and candle burning under controlled laboratory conditions. We also evaluated the performance of the Purple Air monitor against the gravimetric mass measurement and a research level instrument, AQS1 (Aeroqual, New Zealand) in an on-campus residential apartment during several indoor activities including cooking, dishwashing, microwaving, vacuum cleaning, etc. The PurpleAir which showed good performance in the laboratory stage of the project for both candle burning and grilling beef. The correlation coefficients against SMPS were found to be 0.7 and 0.6 for grilling and candle burning, respectively. Also in the residential apartment, the average mass concentrations over 36 hours of continuous operation of Purple Air were 9.7, 14.1, and 15.0 μg/m³ for PM_1, $PM_{2.5}$, and PM_{10}, respectively, while the PM_1, $PM_{2.5}$, and PM_{10} concentrations using the reference gravimetric measurements were 12.7, 20.8, 47.7 μg/m³, respectively. The performance of the Purple Air was better than the AQS1 compared to the reference method in terms of the average mass concentration over 32 hours of the continuous operation. However, when a source like cooking was performed in the apartment, the Purple Air showed the peak concentration with a delay and significantly less than the research level instrument (AQS1). Time and concentration averaging as well as the method of the experiments (locating the monitors near the source or away from the source) can significantly impact the accuracy of the monitors.

Copyright © American Institute of Chemical Engineers. All rights reserved.

598535 The Colorimetric Response of Borohydride Stabilized Silver Nanoparticle on Interaction with Organophosphates

Abstract:

Detection of residual organophosphates in food is of paramount importance in the current scenario. Biosensors for pesticide detection may be developed by monitoring the interaction of nanoparticles with pesticides. In this report, we have studied the colorimetric response of pesticides phorate, chlorpyrifos and malathion with borohydride stabilized silver nanoparticle. The colorimetric changes are corroborated with UV-Visible spectra along with the change in particle size and zeta potential. Nanoparticle on interaction with phorate show visible change in color to dark green while other two pesticides haven't shown any visible color change. The colorimetric response of nanoparticle pesticide mixture was further studied for 12 hours of incubation and it was observed that nanoparticle-phorate mixture started decolorizing after an hour and becomes colorless after 12 hours of incubation, while other two pesticide mixtures remains same. This decolorization of phorate was correlated with the degradation of pesticide in the presence of bare silver nanoparticle. The reaction mechanism was explained by the formation of silver and sulfur atom (Ag→S) complex. The uncapped nanoparticles get stabilized by pesticides over time for chlorpyrifos and malathion while phorate undergoes degradation due to scission of S-CH_2 present in the molecule. This color change may be utilized as a colorimetric method for the detection of phorate pesticide.

Colorimetric Response of Bare Silver Nanoparticle with pesticide

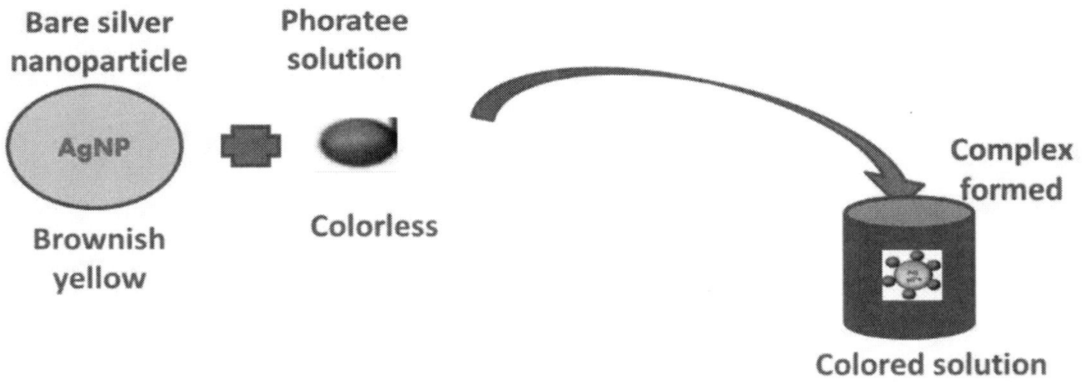

Copyright © American Institute of Chemical Engineers. All rights reserved.

598537 Analysis of Organophosphate Residue Detection on SOLID Surface Using QCM-D As Sensing Platform

Abstract:

Detection of residual organophosphates in food is of paramount importance in current scenario. The understanding of interaction of pesticides on chemically modified surface is the backbone of biosensor fabrication. Quartz crystal microbalance with dissipation is one of the most efficient techniques to analyze the interaction of organic molecules with silane modified surfaces making it desirable bio-sensing platform.

In this report, we have studied the real time adsorption of pesticide phorate and malathion on silica coated quartz crystal of QCM-D. The interaction study was first monitored on gold coated and unmodified silica coated crystal as reference. Further self-assembled monolayer was created on silica crystal using n-propyltrimethoxy silane (PTMS), octyltrimethoxy silane (OTMS), N-(3-(trimethoxysilyl) propyl) aniline and diethoxymethylphenyl silane. The interaction study was also monitored by using AFM analysis. Amount of pesticide adsorbed on surface was evaluated in terms of the mass and thickness change of quartz crystal which is calculated from frequency and dissipation factor change using Sauerbrey equation. Low value of dissipation factor (ΔD) corresponds to stronger binding of pesticide on crystal. The overall results indicates that OTMS modified surfaces contributes mostly to the dissipation changes (ΔD) or acoustic ratio ($\Delta D/\Delta f$). This surface shows largest amount of pesticide adsorption for both malathion and phorate which is correlated with the more hydrophobicity of the surface analyzed by contact angle measurements. The adsorbed amount of pesticide increases with increase in hydrophobicity of surface. The diethoxymethylphenyl silane also shows significant interaction as the phenyl group facilitates interaction with sulfur bond of pesticide. Further verification was done using Raman spectroscopy. The interaction study may be utilized further to fabricate sensors for pesticide detection.

Copyright © American Institute of Chemical Engineers. All rights reserved.

603653 Energy Harvesting Wireless Sensors Using Magnetic Phase Transition

Abstract:

Most industries are connected by cyber-physical systems (CPS), and data and energy in manufacturing technologies are exchanged and shared or sustainable development. To deploy CPS, it is necessary to develop overall security systems, to increase their efficiency including data transfer systems in the network and acquisition systems, and to find new energy sources.

To overcome these energy-related issues of CPS, the possibility of integrating data acquisition sensors with energy harvesting system has been investigated to design energy harvesting sensors in this research. In this energy harvesting system, the magnetic phase transition resulting from changes in temperature is combined with electromagnetic induction, leading to conversion from heat to electricity in the system. In addition, we can successively measure the electromotive force as sensing targets. The proposed system can provide wireless temperature or velocity sensors that directly measure electromotive forces generated by Faraday's law of induction without any additional energy input. Our proposed energy harvesting sensors have the promising potential to contribute significantly to the development of CPS in the near future.

Copyright © American Institute of Chemical Engineers. All rights reserved.

608134 Ultrastretchable Conductive Polymer Complex As a Strain Sensor with a Repeatable Autonomous Self-Healing Ability

Abstract:

Wearable strain sensors are essential for the realization of applications in the broad fields of remote healthcare monitoring, soft robots, and immersive gaming, among many others. These flexible sensors should be comfortably adhered to the skin and capable of monitoring human motions with high accuracy, as well as exhibiting excellent durability. However, it is challenging to develop electronic materials that possess the properties of skin - compliant, elastic, stretchable, and self-healable. This work demonstrates a new regenerative polymer complex composed of poly(2-acrylamido-2-methyl-1-propanesulfonic acid), polyaniline, and phytic acid as a skin-like electronic material. It exhibits ultrahigh stretchability (1935%), repeatable autonomous self-healing ability (repeating healing efficiency >98%), quadratic response to strain ($R^2 > 0.9998$), and linear response to flexion bending ($R^2 > 0.9994$), outperforming current reported wearable strain sensors. The deprotonated polyelectrolyte, multivalent anion, and doped conductive polymer, under ambient conditions, synergistically construct a regenerative dynamic network of polymer complex cross-linked by hydrogen bonds and electrostatic interactions, which enables ultrahigh stretchability and repeatable self-healing. Sensitive strain responsive geometric and piezoresistive mechanisms of the material owing to the homogeneous and viscoelastic nature provide excellent linear responses to omnidirectional tensile strain and bending deformations. Furthermore, this material is scalable and simple to process in an environmentally friendly manner, paving the way for the next-generation flexible electronics.

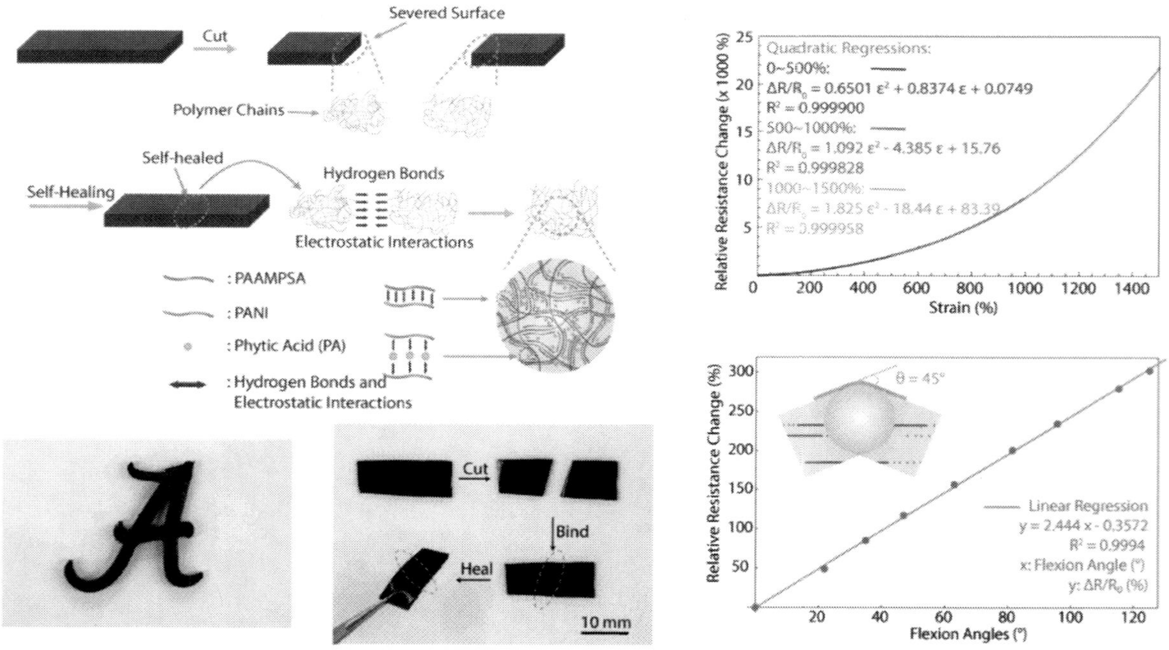

Copyright © American Institute of Chemical Engineers. All rights reserved.

610602 Monitoring Inconsistent ELP Surface Modification By Thiol Desorption

Abstract:

Elastin-like polymers (ELPs) have gained attention recently for their potential as a transduction enhancement for surfaces. However, consistent surface modification is required to ensure a consistent response model. Gold surfaces with ELP has been a challenge in practice, but can be achieved with careful modification procedure, keeping variables such as temperature, ELP concentration, time of modification, and dissolution procedure consistent. We hypothesize that further ELP surface modification inconsistencies are due to ELP that physiosorbs to the gold surface instead of forming new chemisorption modification sites. This presentation will share data demonstrating that inconsistencies in electrochemical measurements of surface-bound ELP are due to inconsistent physisorption of non-chemisorbed ELP molecules. Cyclic voltammetry in KOH was applied to surface-bound ELP at different ELP concentrations in order to reduce the ELP - gold thiol surface. The area under the reduction peak is equal to amount of ELP directly bound to the surface. Electrochemical impedance spectroscopy (EIS) was used with a Randles circuit model to determine the charge transfer resistance of the surfaces. By comparing the trend of charge transfer resistance to surface-bound ELP, we show that optimization coverage points for ELP modified surfaces.

Copyright © American Institute of Chemical Engineers. All rights reserved.

597825 CRISPR-Cas Systems Mediated Electrochemical Biosensing Platforms

Abstract:

Robust developments of personalized medicine for next-generation healthcare highlight the need for sensitive and accurate point-of-care platforms for quantification of disease biomarkers. Broad presentations of clustered regularly interspaced short palindromic repeats (CRISPR) as an accurate gene editing tool indicate that the high-specificity and programmability of CRISPR system can be utilized as a powerful biosensing tool. Herein, we introduce how CRISPR-Cas systems can modulate electrochemical biosensing platforms toward sensitivity, accuracy and programmability. The presented examples demonstrate the potential of CRISPR-Cas systems based biosensing platforms for the development of ideal integrated point-of-care devices.

Copyright © American Institute of Chemical Engineers. All rights reserved.

601915 Synergistic Cancer Gene Therapy Using Multi-Modal Gene Carriers in Combination with Chemotherapy

Abstract:

Bcr-Abl translocation (aka Ph) is a common underlying molecular cause of chronic myeloid leukemia (CML) and other forms of hematological malignancies. In Ph+ leukemia, the pro-apoptotic gene, Bim, is down-regulated, which is further complemented by and the up-regulation of pro-survival Mcl-1. To engineer Ph+ leukemia for facilitated apoptosis, hybrid particles is developed containing an adeno-associated virus (AAV) core expressing Bim and a stimuli-degrading polymeric shell co-delivering MCL-1 siRNA (BIM/MCL-1 hybrid gene carriers). AAV has become a key gene vector with multiple FDA-approved drugs but still offers challenges of immunogenicity and limited capability of genetic modifications. BIM/MCL-1 hybrid gene carriers are developed to avoid immune recognition and also incorporate a second, synergistic therapeutic component to expand the platform's potency. The ketalized polymeric shell that encapsulates the AAV core avoids immune recognition, enhances endosomal escape, and releases the therapeutic payloads in target intracellular destinations. Bim/Mcl-1 hybrid gene carriers were able to synergistically induce apoptosis of Ph+ leukemia via simultaneously up-regulating Bim and silencing Mcl-1, effectively eradicating Ph+ cells in vitro and in vivo. To further synergize the therapeutic efficacy, dasatinib, a tyrosine kinase inhibitor (TKI), was co-administered withBIM/MCL-1 hybrid gene carriers, allowing for nearly full eradication of Ph+ cells, while having minimal effect on Ph- cells as well as reducing the required dasatinib dosage. This talk will cover the synthesis and characterization of Bim/Mcl-1 hybrid gene carriers, and their proof-of-concept demonstration on effective treatment of Ph+ leukemia by themselves and in combination with targeted chemotherapeutics.

Copyright © American Institute of Chemical Engineers. All rights reserved.

603873 Neutrophil Extracellular Traps and Extracellular Vesicles: Mediators of Metastasis in Breast Cancer?

Abstract:

One in eight women will be diagnosed with breast cancer at some point in their lives. In the **United States, the incidence of breast cancer approached 268,600 women in 2019 with almost 42,000 associated deaths**. Approximately 90% of breast cancer mortality is caused from metastases to other vital organs including lungs, liver, brain and bone marrow. Once breast cancer has spread, the 5 year survival rate is only 25%. Thus, understanding the mechanisms by which metastasis occurs is critically important to improving patient outcome. Metastasis has been noted to be facilitated by processes involving the interaction of neutrophils, neutrophil extracellular traps (NETs), and extracellular vesicles (EVs). Indeed, higher concentrations of EVs in breast cancer patient serum as well as increased NET markers are both correlated with poor outcome and increased mortality. NETs (which form via a process called NETosis) occur when neutrophils unravel and extrude their nuclei into webs of DNA that are decorated with granular proteins; NETs have antimicrobial and proinflammatory properties. In the context of cancer, NETs have been shown to aid in arresting circulating tumor cells to promote metastasis, reactivating dormant metastatic tumor cells, and supporting the hypercoagulable state leading to other cancer-associated complications such as thromboembolism. EVs form when submicron membrane-bound particles are released from cells, and function as a mode of intercellular communication; however, their role in promoting NETosis and other metastasis-supporting phenomena remains unclear. In this study, 4T1 murine mammary carcinoma cells were orthotopically injected into n=32 mice, with IVIS performed weekly for 4 weeks to monitor tumor growth. Possible metastases were observed via IVIS in week 2, with gross metastases apparent upon lung collection after 3- and 4-weeks post-tumor injection. Plasma, lung, liver, and primary tumor were collected from groups on a weekly basis, with EVs subsequently isolated from plasma. EVs from each tumor group will be used to stimulate isolated neutrophils from healthy mice *in vitro*. The potential of tumor plasma-derived EVs to induce NETosis will be compared to 4T1 EVs derived from conditioned cell culture medium. Confocal microscopy will evaluate if EVs are fully internalized by neutrophils or bound to their surface. Electron microscopy and NanoSight Tracking Analysis (NTA) will be utilized to validate morphological differences in EV populations. Finally, western blots will be performed to validate the presence of EV-associated markers (CD9 and ALIX), tumor-specific EVs (EpCAM), and tissue factor, which can lead to hypercoagulability and NETosis downstream of platelet activation. We hypothesize that EVs from timepoints correlating with the onset of common metastases to lung and liver (week 2 and 3) will display enhanced potential to cause NETosis compared to EVs from timepoints closer to primary tumor establishment (week 1) or EVs in mice with later-stage metastases (week 4 and beyond).

Copyright © American Institute of Chemical Engineers. All rights reserved.

603497 Understanding the Roles of Lean and Obese Adipose Tissue on Triple Negative Breast Cancer Recurrence

Abstract:

Background: Triple negative breast cancer (TNBC) patients have poor prognoses, high locoregional recurrence rates, and low post-recurrence survival rates, and these statistics are exacerbated for patients with obesity. Our group's previous work in a pre-clinical model has shown that recurrence in TNBC may be facilitated by circulating tumor cells (CTCs) recruited to the primary tumor site following radiation therapy due to chemotactic cues from the damaged normal tissue. Our current work explores how specific cells within adipose tissue, such as adipocytes, respond to radiation damage, how these cells impact immune cell recruitment and the wound healing process, and how these processes change in the context of obesity and impact TNBC recurrence.

Methods: For *in vitro* experiments, murine 3T3-L1 pre-adipocytes were differentiated into adipocytes, which were then exposed to 10 Gy of ionizing radiation (IR) and evaluated against control cells receiving no radiation. Conditioned media (CM) from adipocytes was collected at 2 and 7 days post-IR and utilized in proliferation and invasion assays. Luciferase-labeled 4T1 cells were exposed to adipocyte CM for 48 hours and proliferation was determined from bioluminescence measurements. For co-culture proliferation studies, luciferase-labeled 4T1 cells were co-cultured with irradiated adipocytes for 48 hours. Additionally, 4T1 cells were used in a transwell invasion assay with CM from irradiated adipocytes as the chemoattractant.

For *in vivo* experiments, 6-week-old female BALB/c mice were fed a high-fat diet (60% kcal from fat) to induce obesity and were compared to control groups fed a sucrose-matched low-fat diet (10% kcal from fat) for 15 weeks. The inguinal mammary fat pads (MFPs) were resected for further analysis. MFPs from both diet-induced obesity (DIO) and lean groups were subjected to either 0 or 20 Gy of *ex vivo* IR and incubated for 2 days in serum-free media for CM collection. MFPs were then fixed, processed, and stained for hematoxylin and eosin (H&E) analysis. Cytokine expression in CM between the groups was determined via a murine 31-plex Luminex assay.

Results: 4T1 cells exposed to CM collected from adipocytes 7 days post-IR showed a significant increase in proliferation when compared to that of control CM ($p < 0.05$), which was confirmed in co-culture studies. This result was interesting especially since there was no difference in proliferation between TNBC cells exposed to CM from either irradiated or control adipocytes at the 2-day time point. Additionally, an increase in migration, but not invasion, was observed when TNBC cells were exposed to CM from radiation-damaged adipocytes at 2 days post-IR.

Obese mice showed an over 2-fold increase in MFP weight ($p < 0.001$) and microvesicular steatosis compared to lean mice. After *ex vivo* radiation, MFPs from DIO mice exhibited differences in adipocyte morphology, including decreased density and variable size, and decreased ECM deposition compared to lean MFPs, suggesting that IR promotes distinct microenvironmental changes in obese tissues. Cytokine analysis of CM collected from irradiated DIO compared to lean MFPs showed a greater than 2-fold upregulation in post-IR secretion of CXCL-2, CXCL-10, and IL-6 and an almost 2-fold downregulation of IL-5 and IL-7 secretion. This altered secretion profile highlights the unique radiation damage response in obese MFPs and suggests that irradiated obese MFPs may influence immune cell recruitment dynamics, which will be studied in future experiments.

Conclusions: Our study demonstrates the potential role of radiation-damaged adipocytes and obese mammary adipose tissue in the recruitment and proliferation of CTCs following radiotherapy. This work underscores the importance of evaluating individual cell and whole tissue responses to radiation damage in the microenvironment as a consequence of primary tumor treatment to better understand how wound healing impacts recurrence. These results suggest that the damaged stroma may facilitate tumor regrowth, which merits further exploration and could lead to new insights into TNBC recurrence mechanisms.

Copyright © American Institute of Chemical Engineers. All rights reserved.

607291 Palmitate Induced Desmoplakin Loss and Enhanced Migration of Breast Cancer Cell Mediated By the IRE1α-XBP1-ZEB1 Pathway

Abstract:

Elevated levels of the saturated fatty acid, palmitate (PA), have been linked to metastatic progression of cancer cells; however, the precise mechanism behind this connection is not well understood. The loss of cell adhesion proteins, such as desmoplakin (DSP), plays an important role in driving the transformation events of cancer cells to become metastatic. Our laboratory previously showed PA activated the IRE1α-XBP1 (inositol-requiring enzyme 1α-X-box binding protein 1) pathway, which is a branch of the endoplasmic reticulum (ER) stress response. Here, we investigated the role PA plays in the activation of the IRE1α-XBP1 pathway in promoting the loss of DSP in breast cancer cells. We found that treating a human breast cancer cell line, MDA-MB-231, with PA induced the loss of DSP and enhanced cell migration mediated by the IRE1α-XBP1 pathway. Furthermore, silencing DSP was sufficient to promote cell migration. We found that zinc finger E-box binding homeobox 1 (ZEB1) transcription factor functions as a direct transcriptional repressor of DSP and that PA increased the expression of ZEB1. CRISPR-Cas9 knockout of IRE1α abolished the PA-induced increase in ZEB1 expression, decrease in DSP levels, and cell migration in the breast cancer cells. Based on these findings, we propose that the loss of DSP and the increase in cell migration in breast cancer cells induced by PA is mediated by the IRE1α-XBP1-ZEB1 pathway.

Copyright © American Institute of Chemical Engineers. All rights reserved.

607366 Polymerized Hemoglobin Facilitated Modulation of Tumor Oxygenation Is Dependent on Tumor Oxygenation Status and Oxygen Affinity of the Hemoglobin-Based Oxygen Carrier

Abstract:

Background: Hemoglobin (Hb) based oxygen (O_2) carriers (HBOCs) are a promising class of O_2 therapeutic that can relieve chronic hypoxia in the tumor microenvironment, which may increase the effects of chemotherapy. Previous studies have demonstrated that HBOC O_2 affinity significantly affects how HBOC treated tumors respond to chemotherapy. However, there is no consensus concerning whether high or low O_2 affinity HBOCs results in greater chemosensitization. To address these concerns, we investigated how a low-O_2-affinity tense (T) quaternary state polymerized hemoglobin (PolyHb) and a high-O_2-affinity relaxed (R) quaternary state PolyHb impact tumor oxygenation and sensitivity to chemotherapeutics. These effects were evaluated with a combination of *in silico* and *in vivo* methods.

Methodology: For this study, we synthesized a library of low molecular weight (MW) PolyHb in the T- and R-state. The following biophysical properties of these materials were measured: O_2 affinity, gaseous ligand binding kinetics, MW distribution, and hydrodynamic diameter. With these properties, we simulated O_2 transport within a three dimensional vascularized tumor construct. After simulating the tumor, we used principal component analysis to examine which properties influence tumor oxygenation. We examined how the T-state and R-state PolyHb modulates O_2 delivery to hypoxic and normoxic human melanoma xenografts in a murine window chamber model. We examined microcirculatory fluid flow via video shearing optical microscopy, and oxygen distributions via phosphorescence quenching microscopy. Additionally, we examined how weekly infusion of a 20% top-load dose of PolyHb influences tumor growth rate, vascularization, and regional blood flow. To test the chemosensitizing potential, we administered an R-state PolyHb to three different non-small cell lung cancer cell lines cultured in vitro and implanted *in vivo* into healthy mice.

Results and Conclusions: In general, we found that infusion of T-state PolyhHb is more likely to decrease tissue hypoxia and modulate the metabolic rate of O_2 consumption. Infusion of T-state and R-state PolyHbs led to worse outcomes in normoxic tumors. Of particular concern was the high-affinity R-state PolyHb, which led to no improvement in hypoxic tumors and significantly worsened outcomes in normoxic tumors. Taken together, the results of this study indicate that the tumor oxygen status is a primary determinant of the potency and outcomes of infused PolyhHb. The R-state PolyHb administered in this study is unable to deliver O_2 unless under severe hypoxia, which significantly limits its oxygenation potential. *In vitro* sensitivity studies indicate that the administration of PolyHb increases the effectiveness of cisplatin under hypoxic conditions. Additional animal studies revealed that co-administration of R-state PolyHb with cisplatin attenuated tumor growth without alleviating hypoxia. Analysis of reactive O_2 species production in the presence of hypoxic culture indicates that exogenous ROS production by oxidized PolyHb may provide the mechanism of chemosensitization. This ROS mechanism, coupled with oxygenation, may be a potential chemosensitizing strategy for use in cancer treatment.

Copyright © American Institute of Chemical Engineers. All rights reserved.

610812 Tumor Acidic Microenvironment Impairs Cancer Cell Migration

Abstract:

One of the key features of tumors is their acidic microenvironment[1]. Although in normal tissue, pH is tightly regulated, tumor cells, which have very high proliferation rates, create an acidic microenvironment by secreting acidic materials due to high glycolytic activity (Warburg Effect) and due to poor perfusion of the dense tumor tissue[1]. Cell migration is a key step in the process of cancer metastasis. Metastasizing cells travel through confining 3D pores, fiber or channel like tracks[2]. Previous studies using 2D and Transwell assays have reported conflicting results showing different trends about the effects of extracellular acidosis on cancer cell migration and invasion[3,4,5,6]. Thus, we employed array of sophisticated in vitro models to delineate the role of acidosis in key steps of metastatic cascade.

In this study, we employed PDMS-based microfluidic devices to track cell migration inside channels of different dimensions. We also studied 2D migration as well as tumor cell dissociation from 3D breast cancer organoids. In separate experiments, cells were seeded on a 2D area between two sets of 1 mm long microchannels leading to either acidic or normal local microenvironment (Fig. 1A). The effects of acidosis were also investigated using mouse models.

We measured reduced cell motility in 2D and inside microchannels under acidosis. The most pronounced differences were observed at an extracellular pH of 6.4. This pH also interfered with tumor cell dissociation from 3D breast cancer organoids (Fig. 1C). We determined that the reduced migratory capacity of cells grown in an acidic microenvironment is due to decreased mRNA and protein expression of Na^+/H^+ exchanger, NHE1. Along these lines, we determined that NHE1 activity is also significantly lower under acidosis, as measured by the NH_4Cl pre-pulse technique. This is further substantiated by data showing that NHE1 knockdown in cells grown at the pH of 7.4 suppresses motility down to the levels with cells cultured at pH 6.4 for 24 hours. In line with our in vitro findings, breast cancer cells, pre-exposed to acidic pH prior to their injection in the tail vein of mice, exhibited reduced levels of lung colonization. Interestingly, cells seeded on 2D preferentially moved through microchannels toward physiological pH of 7.4 as opposed to pH 6.4 in significantly larger numbers and with faster speeds (Fig. 1B). The tendency of cells to move away from an acidic microenvironment was observed for breast cancer cells that were preconditioned for 24 hours at either pH_e 7.4 or 6.4. This result may explain tumor cells' propensity to leave the primary tumor during metastasis.

We propose that extracellular acidosis impairs cancer cell migration on 2D and in 3D-like microfluidic channels as well as cell dissociation from breast cancer organoids, presumably by downregulating both the expressions and activity of NHE1. Although acidosis impairs cell migration, we posit that cancer metastasis is in part driven by tumor cells' propensity to migrate away from the acidic tumor environment they create towards physiological pH environments. Our study expands our understanding of cancer cell behavior in acidic tumor microenvironments, which could aid in future cancer research and development of cancer therapeutics.

References

1. S.K. Parks et. al. Nat Rev Cancer, (2013) 13, 611-623.

2. C.D. Paul et. al. Nat. Rev. Cancer (2017) 17,131–140 .

3. S.C. Gupta et. al. Oncotarget, (2014) 5:12070-12082.

4. Y. Wu et. al. Tumor Biol, (2017) Jun; 39(6):1010428317705750.

5. C. Chen et. al., Exp Cell Res, (2010) Oct 15;316(17):2910-21

6. R.D. Castellone et. al., Cancer Letter, (2011)

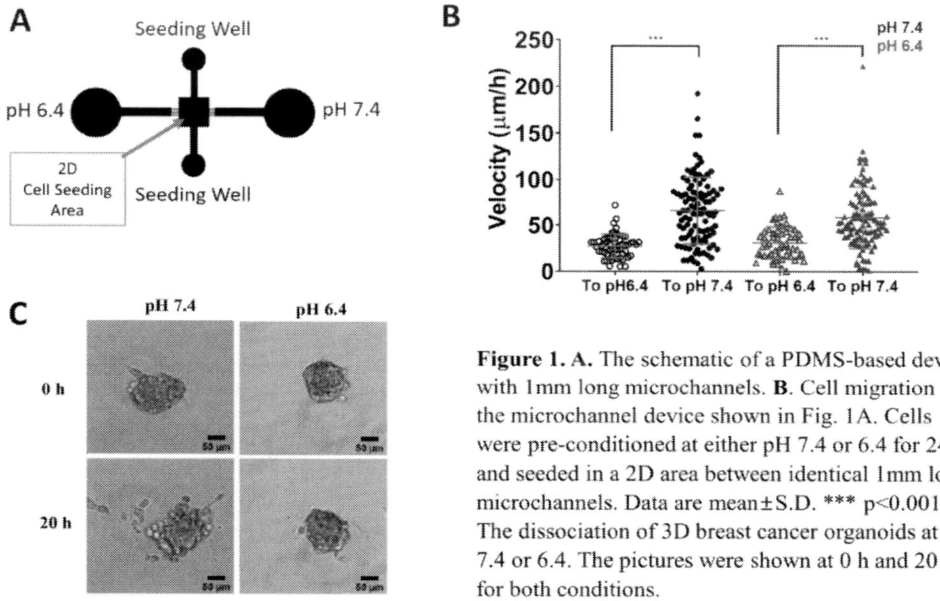

Figure 1. A. The schematic of a PDMS-based device with 1mm long microchannels. **B**. Cell migration in the microchannel device shown in Fig. 1A. Cells were pre-conditioned at either pH 7.4 or 6.4 for 24h and seeded in a 2D area between identical 1mm long microchannels. Data are mean±S.D. *** p<0.001 **C**. The dissociation of 3D breast cancer organoids at pH_e 7.4 or 6.4. The pictures were shown at 0 h and 20 h for both conditions.

Copyright © American Institute of Chemical Engineers. All rights reserved.

608077 Normalizing the Tumor Microenvironment for Immunotherapy with Dexamethasone and Tranilast

Abstract:

Cancer cells promote the development of abnormal blood and lymphatic vessels through altered signaling and generation of mechanical stress. These processes create a hostile tumor microenvironment featuring hypoxia, acidity, and elevated interstitial fluid pressure. These abnormalities fuel disease progression, immunosuppression, and treatment resistance[1]. Judicious use of antiangiogenesis agents that block altered vascular signaling can transiently 'normalize' tumor vessels[2]. In parallel, because desmoplasia collapses vessels, normalizing the extracellular matrix to decompress vessels also can improve their function[3]. These two strategies may be combined to enhance chemo-, radio-, and immune-therapies[4].

Here, I will describe our progress in repurposing commonly used drugs to normalize the tumor microenvironment to increase efficacy of immunotherapy[5]. In one approach, we repurposed the anti-hypertensive drug losartan to normalize the extracellular matrix, alleviate hypoxia, and potentiate chemotherapy[6]. Losartan is efficacious in pancreatic cancer patients receiving chemo-radiation[7]and is being tested with immunotherapy (NCT03563248). Because anti-hypertensives might not be safe in hypotensive patients, we reformulated this class of anti-hypertensive drug in a nanocarrier to reduce its intended effect[8].

In a second approach, we repurposed two drugs already administered to most cancer patients – glucocorticoid steroids and anti-histamines. Using dexamethasone as a model glucocorticoid steroid, we found that a certain dose and schedule can normalize vessels and the extracellular matrix thereby alleviating hypoxia and increasing nanocarrier delivery and efficacy[9]. In parallel, we found that a certain anti-histamine (i.e. tranilast) normalizes the extracellular matrix towards decompressing vessels and increasing nanocarrier and immune checkpoint inhibitor efficacy[10]. Outstanding questions regarding combining glucocorticoid steroids with immune checkpoint inhibitors and overcoming resistance in breast cancer lung metastasis will be addressed.

Repurposing drugs to normalize the tumor microenvironment has the potential to rapidly and inexpensively expand the fraction of cancer patients that can benefit from immunotherapy.

References:

1. Martin, J. D.; Fukumura, D.; Duda, D. G.; Boucher, Y.; Jain, R. K., Reengineering the Tumor Microenvironment to Alleviate Hypoxia and Overcome Cancer Heterogeneity. *Cold Spring Harbor Perspectives in Medicine* **2016,***6*, a027094.
2. Chauhan, V. P.; Stylianopoulos, T.; Martin, J. D.; Popovic, Z.; Chen, O.; Kamoun, W. S.; Bawendi, M. G.; Fukumura, D.; Jain, R. K., Normalization of Tumour Blood Vessels Improves the Delivery of Nanomedicines in a Size-Dependent Manner. *Nat Nanotechnol* **2012,***7*, 383-388.
3. Stylianopoulos, T.; Martin, J. D.; Chauhan, V. P.; Jain, S. R.; Diop-Frimpong, B.; Bardeesy, N.; Smith, B. L.; Ferrone, C. R.; Hornicek, F. J.; Boucher, Y.; Munn, L. L.; Jain, R. K., Causes, Consequences, and Remedies for Growth-Induced Solid Stress in Murine and Human Tumors.

Proc. Natl. Acad. Sci. U. S. A. **2012,** *109*, 15101-15108.
4. Martin, J. D.; Seano, G.; Jain, R. K., Normalizing Function of Tumor Vessels: Progress, Opportunities and Challenges. *Annual review of physiology* **2019,** *81*, 505-534.
5. Martin, J. D.; Cabral, H.; Stylianopoulos, T.; Jain, R. K., Improving Cancer Immunotherapy Using Nanomedicine: Progress, Opportunities and Challenges. *Nat. Rev. Clin. Oncol.* **2020,** *17*, 251-266.
6. Chauhan, V. P.; Martin, J. D.; Liu, H.; Lacorre, D. A.; Jain, S. R.; Kozin, S. V.; Stylianopoulos, T.; Mousa, A. S.; Han, X.; Adstamongkonkul, P.; Popovic, Z.; Huang, P.; Bawendi, M. G.; Boucher, Y.; Jain, R. K., Angiotensin Inhibition Enhances Drug Delivery and Potentiates Chemotherapy by Decompressing Tumour Blood Vessels. *Nat Commun* **2013,** *4*, 2516.
7. Murphy, J. E.; Wo, J. Y.-L.; Ryan, D. P.; Clark, J. W.; Jiang, W.; Yeap, B. Y.; Drapek, L. C.; Ly, L.; Baglini, C. V.; Blaszkowsky, L.; Ferrone, C.; Parikh, A. R.; Weekes, C.; Nipp, R. D.; Kwak, E. L.; Allen, J. N.; Corcoran, R. B.; Ting, D. T.; Faris, J. E.; Zhu, A. X., et al., A Phase Ii Study of Neoadjuvant Folfirinox in Combination with Losartan Followed by Chemoradiotherapy in Locally Advanced Pancreatic Cancer: R0 Resection Rate and Clinical Outcomes. *JAMA Oncology* **2019,** *5*, 1020-1027.
8. Chauhan, V. P.; Chen, I. X.; Tong, R. T.; Ng, M. R.; Martin, J. D.; Naxerova, K.; Wu, M. W.; Huang, P.; Boucher, Y.; Kohane, D. S.; Langer, R.; Jain, R. K., Reprogramming the Microenvironment with Tumor-Selective Angiotensin Blockers Enhances Cancer Immunotherapy. *Proc. Natl. Acad. Sci. U. S. A.* **2019,** *116*, 10674-10680.
9. Martin, J. D.; Panagi, M.; Wang, C.; Khan, T. T.; Martin, M. R.; Voutouri, C.; Toh, K.; Papageorgis, P.; Mpekris, F.; Polydorou, C.; Ishii, G.; Takahashi, S.; Gotohda, N.; Suzuki, T.; Wilhelm, M. E.; Melo, V. A.; Quader, S.; Norimatsu, J.; Lanning, R. M.; Kojima, M., et al., Dexamethasone Increases Cisplatin-Loaded Nanocarrier Delivery and Efficacy in Metastatic Breast Cancer by Normalizing the Tumor Microenvironment. *ACS Nano* **2019,** *13*, 6396-6408.
10. Panagi, M.; Voutouri, C.; Mpekris, F.; Papageorgis, P.; Martin, M. R.; Martin, J. D.; Polydorou, C.; Louca, M.; Kataoka, K.; Cabral, H.; Stylianopoulos, T., Tgf-B Inhibition Combined with Cytotoxic Nanomedicine Normalizes Triple Negative Breast Cancer Microenvironment Towards Anti-Tumor Immunity. *Theranostics* **2020,** *in press*.

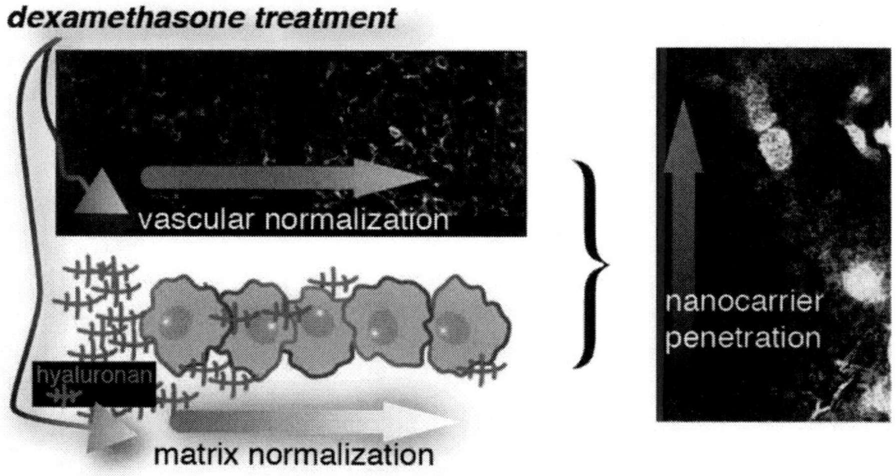

Copyright © American Institute of Chemical Engineers. All rights reserved.

601823 Long Term Tracking of Cell Lineages with Quantitative Phase Imaging

Abstract:

Cell mass is a biophysical parameter which is useful in monitoring cell growth, cell drug response, and mass partition during cell division. We have developed a method which allows us to continuously track the mass of individual cells in culture for up to 5 generations. The principle of cell mass measurement is quantitative phase imaging (QPI). QPI measures the phase shift of light passing through cells and uses the known relationship between phase shift and cell mass to calculate cell mass. For long term cell mass tracking, we perform cell culture and imaging using an array of microwells, 155 µm square x 40 µm high wells that prevent cells from moving out of the field of view during imaging. We fabricated the microwell array using MY-133, a material with the same refractive index as water, to avoid phase unwrapping during QPI data collection. We design the size of individual microwell to fit it into the field of view of our microscope. During the experiment, the microscope scans through multiple microwells in several minutes, then repeatedly scans the same microwells until the dividing cells cover the entire microwell area. We process the images by using an image processing workflow we developed to remove the microwells and noise from the background. Finally, we segment out each cell in the images, calculate their mass, and generate its profile of mass over time. Mass tracking with 70Z/3 B lymphoblast and MCF-7 breast cancer cells by this method shows it can be used for both non-adherent and adherent cells. Our automated results match the expected cell growth rate and can be used to study cell to cell variation within and across lineages. This method has applications in long-term monitoring of cancer cell drug response as well as the study of B cell fate during differentiation to antibody-secreting plasma cells in response to infectious disease.

Copyright © American Institute of Chemical Engineers. All rights reserved.

598126 Cartilage Production in a Novel Catechin-Loaded Poly(N-vinylcaprolactam)-Hyaluronic Acid Hydrogel

Abstract:

Osteoarthritis is a degenerative disease of the articular cartilage lining moving joints that affects approximately 30 million Americans and is a leading cause of disability. There is currently no cure for osteoarthritis, and existing treatment protocols seek to manage pain or replace diseased tissue with non-native materials. Cartilage tissue engineering has emerged as a promising strategy to develop new treatment options; however, this process is far from optimized. The selection of polymers to create scaffolds to encapsulate chondrocytes, the cells contained in cartilage, is a growing area of research as the polymer choice has been shown to affect the mechanical integrity of the engineered tissue, cell viability, and the production of cartilage specific proteins such as collagen (COL) and glycosaminoglycans (GAG). Further, it has been shown that the incorporation of cartilage-specific polymers such as hyaluronic acid, collagen, and others into the scaffold matrix can induce upregulation of extracellular matrix protein production. In this study, we seek to explore the potential of a new thermosensitive hydrogel, PVCL-HA, containing the synthetic polymer poly(N-vinylcaprolactam), covalently bound hyaluronic acid, and catechin, a naturally occurring anti-inflammatory compound. We expect to present preliminary results for osteoarthritic human chondrocytes cultured in three systems: micromass, agarose hydrogels, or PVCL-HA hydrogels. The levels of COL, GAG, and nitric oxide, an indicator of tissue disease and inflammation, will be compared across the three culture systems. We hypothesize that cartilage-specific protein production will be highest for chondrocytes cultured in PVCL-HA hydrogels due to the incorporation of hyaluronic acid in the scaffold. Compared to agarose, PVCL-HA gels provide a culture environment that more closely mimics native cartilage. Further, we hypothesize that nitric oxide levels will be reduced in PVCL-HA hydrogels due to the incorporation of catechin, which has been shown as a potent anti-inflammatory. Should PVCL-HA promote significant improvement in the quality of cartilage tissue, it is expected to prove a viable scaffold option for the production of engineered tissue and the treatment of osteoarthritis.

Copyright © American Institute of Chemical Engineers. All rights reserved.

603198 Engineering of Pancreatic Beta-Cells for Optogenetic Modulation of Insulin Function

Abstract:

Diabetes is characterized by elevated blood glucose (BG) due to autoimmune destruction of insulin-producing β-cells (type 1; T1D) or extensive β-cell apoptosis owed to insulin resistance (type 2; T2D). Over 9% of the US population suffers from diabetes and its complications including blindness, kidney failure and stroke making this the single most expensive disease with total costs of $327 billion in 2017. Pharmacological interventions are linked to side effects and target molecules with low tissue. Daily administration of insulin is essential for the survival of T1D and advanced T2D patients but imposes significant lifestyle restrictions, requires patient compliance, may cause hypoglycemia, and does not reverse the long-term crippling effects of the disease. Therefore, novel strategies to achieve BG homeostasis in a glucose-dependent fashion without side effects are highly desirable.

Optogenetic technologies allow the drug-free manipulation of diverse cellular functions through modulation of molecular moieties with unique spatiotemporal precision. One such moiety is cyclic AMP (cAMP), which is an amplifier of glucose-stimulated insulin secretion (GSIS) in β-cells. To this end, we engineered pancreatic β-cells and primary islets to express a photoactivatable adenylyl cyclase (PAC) [1]. PAC-expressing cells exhibited a greater than five times rise in cAMP within 5 min of photoactivation and a rapid drop (~12 min) upon termination of illumination. This resulted in a 2- to 3-fold increase of GSIS in PAC-positive β-cells or islets exposed to blue light vs. those kept in dark. Cells exhibited consistently enhanced GSIS over multiple rounds of photoactivation without changes in viability. The pronounced response was comparable to that by β-cells treated with known secretagogues including adenylyl cyclase activators or phosphodiesterase inhibitors [1]. No difference was observed in insulin release with or without illumination in the absence of glucose, further justifying the optogenetic targeting of cAMP, which augments but does not induce hormone secretion. The dependence of GSIS on Ca^{2+} signaling remained intact in engineered β-cells. Light-stimulated secretion of insulin was an order of magnitude greater when PAC-expressing β-cells formed clusters ('pseudoislets'; PIs). The function of PIs carrying PAC was tested *in vivo* [2]. Encapsulated PIs were delivered subcutaneously to mice, which were rendered diabetic upon treatment with streptozotocin. The animals were subjected to a glucose tolerance test (GTT) following a 6-h fasting and intraperitoneal injection of glucose. Blood was sampled before (0 min) and after glucose injection (30, 60, 90 and 120 min) and BG was determined using a glucometer. The mice were exposed to blue light for 3 h starting 1 h before the GTT. Animals receiving PAC-expressing β-cells exhibited impaired glucose tolerance in the absence of illumination while the BG returned to baseline in mice with engineered cells undergoing photostimulation.

Currently, we are exploring the use of a PAC that is engineered for activation by near-infrared window light [3]. Red light exhibits greater depth of penetration into tissues. This suggests that a cell delivery design can be realized without the need of integrating the light source for optogenetic stimulation thereby reducing the device complexity and size. Our work will enable technologies combining cellular engineering and optogenetics for the treatment of diabetes.

Acknowledgements:

This work has been partially supported by a grant (CBET-1951104) to ET by the National Science Foundation (NSF).

References:

1. Zhang, E.S. Tzanakakis, Optogenetic regulation of insulin secretion in pancreatic beta-cells, *Sci. Rep.*, 7(1):9357, 2017.
2. Zhang, E.S. Tzanakakis, Amelioration of diabetes in a murine model upon transplantation of pancreatic beta-cells with optogenetic control of cyclic AMP, *ACS Synth. Biol.*, 8:2248, 2019.
3. Fomicheva, C. Zhou, Q-Q Sun, M. Gomelsky, Engineering adenylate cyclase activated by near-infrared window light for mammalian optogenetic applications, *ACS Synth. Biol.*, 8:1314, 2019.

Copyright © American Institute of Chemical Engineers. All rights reserved.

Convection-driven intravascular bioartificial pancreas (iBAP) enhances solute delivery to macroencapsulated human islets.

Ana G. Santandreu[a], Alonso Torres[a], Parsa Taheri-Tehrani[a], Charles Blaha[a,b], Benjamin Feinberg[a], Nathan Wright[a,b], Shuvo Roy[a,b*]

[a]Department of Bioengineering and Therapeutic Sciences, University of California, San Francisco, CA, 94158, USA.
[b]Silicon Kidney LLC, San Francisco, CA, 94143, USA.

Keywords: T1D, Intravascular Bioartificial Pancreas, Silicon Nanopore Membranes, GSIS, Convective Mass Transport.

Introduction: Advances in islet macroencapsulation for the treatment of type 1 diabetes (T1D) have led to increased interest in the development of bioartificial pancreas devices. Many of these devices exhibit limited islet viability and poor glycemic control due to the large diffusion distances of key solutes, such as glucose, oxygen, and insulin, across the islet tissue. Inefficient nutrient delivery rates impair the glucose-insulin kinetics and may even lead to suppression of insulin production due to the lack of oxygen. To maximize nutrient delivery and achieve more efficient mass transfer rates, our lab is engineering a convection-driven intravascular bioartificial pancreas (iBAP) enabled by ultra-high-flux silicon nanopore membranes (SNM). These membranes are produced using standard microfabrication processes that for allow precise control of nanopore size and membrane thickness to achieve high solute delivery rates. We hypothesized that the SNM-based iBAP device provides enhanced oxygenation to support adult human islets in retaining their function. In this work, we examined the oxygen consumption profiles for SNM-generated ultrafiltration rates to islet equivalent doses (UF/IEQ) in the range of 14 – 200 nL/min/IEQ. (1 IEQ = 1 150μm diameter islet). The oxygen profiles were determined with a Finite Element Method (FEM)-based model with atmospheric (160 mmHg) oxygen levels to mimic *in vitro* oxygen levels as the starting concentration. Then, these rates were experimentally tested in a mock-loop *in vitro* set-up for glucose-stimulated insulin secretion (GSIS).

Materials and Methods:

Oxygen modeling: The oxygen model used COMSOL Multiphysics to define our system's geometry (Figure **1.A.**), which includes convective flow through the scaffold's microchannels and diffusive transport with reactive solute rates in the islet tissue. The post-processing cut-line feature in COMSOL was used to determine the worst-case oxygen profile scenario of each condition; i.e., the islets farthest away from the hexagonally arranged microchannels. The Navier-Stokes and mass continuity equations for incompressible Newtonian fluids were used to describe the velocity field due to convection, whereas consumption rates were assumed to follow Michaelis-Menten kinetics and dependent on the glucose concentration. This model established the oxygen consumption profiles of islet scaffolds at 14, 28, 56, 100 and 200 nL/min/IEQ.

In vitro set-up: The SNM-based iBAP (Figure **1.B.**) was composed of a polycarbonate blood flow-path sandwiched by an SNM-islet chamber stack on each side. The CNC machined 316L stainless-steel islet chambers (Prototek Manufacturing Services, Inc. (Sunnyvale, CA)) held the adult human islets (UCSF Islet and Cellular Production Facility (San Francisco, CA) and Prodo Laboratories, Inc (Aliso Viejo, CA)) in a 3% (w/v) ultra-low gelling agarose scaffold (Sigma: 9012-36-6) patterned with microchannels (Figure.**1.C.**). Each islet chamber was sealed with a polycarbonate backside containing an ultrafiltrate (UF) outlet. A dynamic GSIS assay was performed by connecting the iBAP device to a mock circulatory flow system conditioned at 37°C and 5% CO_2. The encased islets were stabilized in a 2-hour continuous flow period represented as the "pre-stimulatory phase" at 5 mM of glucose (L1). Ultrafiltrate samples were collected for the last 16 minutes of this period. The glucose concentration was then increased to 28 mM (H) and samples were collected for 30 minutes. Lastly, the glucose concentration was reduced to 5 mM (L2) and samples were collected for 32 minutes. The insulin content in the UF samples was quantified via an enzyme-linked immunosorbent assay (Mercodia: 10-1113-01).

Results and Discussion: The oxygen model revealed that perifusion rates as low as 28 nL/min/IEQ provided sufficient oxygen to maintain insulin production, which requires oxygen tensions above the cut-off threshold value of 25 mmHg (or 0.034 mol/m^3) to avoid inhibition of insulin secretion. In addition, GSIS curves of rates above 28 nL/min/IEQ exhibited the characteristic biphasic secretion pattern generally observed with human islets in traditional perifusion systems (Figure **1.D.**) and included a shutdown of insulin production upon transition to low glucose (L2) medium. Therefore, the correlation between oxygen profile and GSIS kinetics for the perifusion rates of 14 and 28 nL/min/IEQ validate the oxygen model, given that 14 nL/min/IEQ exhibited poor glucose-insulin kinetics. Interestingly, the 100 nL/min/IEQ condition averaged a higher stimulation index (13.86±7.29) compared to the higher 200 nL/min/IEQ ratio tested (5.30±2.48), suggesting that higher perifusion rates may affect insulin production possibly due to damage induced from shear forces at the periphery of the islets exposed to higher flows.

Conclusion: Human islets in the iBAP device matched the GSIS pattern observed with standard islet perifusion studies, indicating the islet scaffold can ensure rapid glucose-insulin kinetics under convective mode. Furthermore, a clinically relevant insulin production was observed with a relatively small density of islets *in vitro* since the most promising perifusion rates of 100 and 200 nL/min/IEQ correlated to islet densities as little as 2.5% volume of IEQs (~500 IEQs) per volume of a 36-µL islet scaffold. Therefore, this work will inform future islet dosing experiments and device size necessary to systemically deliver insulin to treat T1D.

Acknowledgements: This research was supported by JDRF (Grant #: 3-SRA-015-37-Q-R) and the NIH/NIDDK Small Business Innovation Research (SBIR) Program (Project #: 5R44DK104299-03). We would also like to thank colleagues Drs. Rachel Gurlin and Rebecca Gologorsky for their insightful contributions.

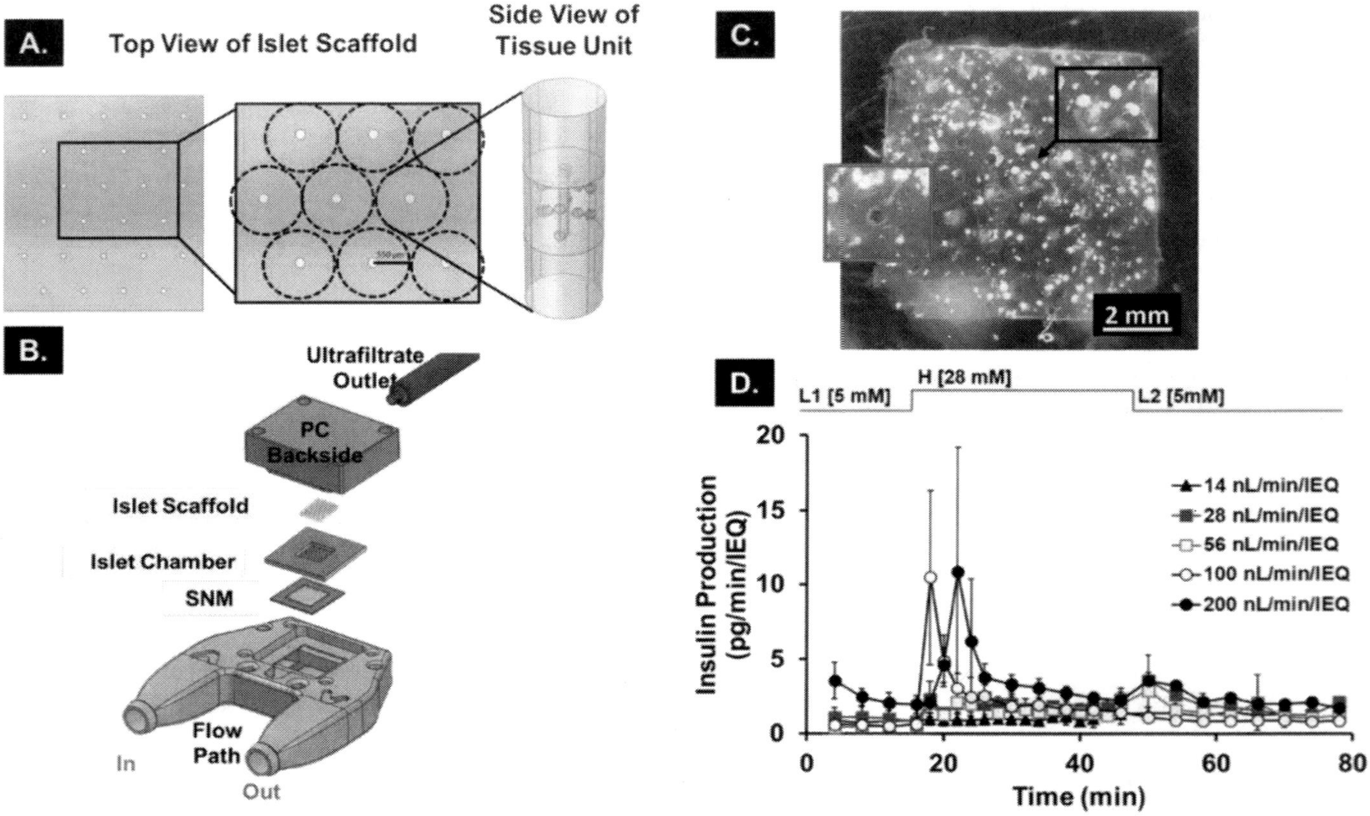

Figure 1. ***in vitro* testing of adult human islets in a convection-driven iBAP device enabled by silicon nanopore membranes (SNMs) – A:** Top view of the islet scaffold design with hexagonal arrangement of 150-µm diameter microchannels (white circles) enclosed in the tissue unit of 550 µm radius (segmented black circles), which was defined by the symmetry boundary condition in COMSOL. **B.** Exploded view of the iBAP device assembly. **C.** Dissecting microscope image of 36-µL islet scaffold with 56 hexagonally arranged microchannels (red box) and encapsulated adult human islets black box). **D.** Glucose-stimulated insulin secretion (GSIS) curves of SNM-generated islet perifusion rates ranging from 14 – 200 nL/min/IEQ as a function of time.

611084 Stretching and Gallic Acid Promote Extracellular Matrix Production in Human Articular Chondrocytes

Abstract:

Articular Cartilage (AC) is an avascular protective soft tissue that covers bones in articulating joints. The avascular nature of AC makes its access to oxygen and nutrients limited and reduces its ability to regenerate after trauma. AC degenerates gradually after injury leading to osteoarthritis (OA), the most common form of arthritis worldwide. Although OA was categorized as "non-inflammatory" for a long time in contrast to rheumatoid arthritis, recent studies reconsidered the definition after observing immune responses and apparent symptoms of inflammation, e.g., swelling, in the joints of OA patients. Another critical symptom of inflammation is the upregulation of catabolic genes that destroy the extracellular matrix (ECM) proteins of cartilage, making cartilage mechanically weaker. In this work, we studied the use of of biochemical and mechanical factors for reducing the inflammation of human articular cartilage cells, chondrocytes (hAChs). To do that, human chondrocytes were pooled, taken from four different OA study participant tissues removed after total knee arthroplasty. hACHs were cultured within two separate groups, a static group and a group mechanically stimulated by stretching. Group samples were either treated with standard culture medium as a negative control or aa nutraceutical supplemented medium as the experimental group. Nutraceuticals were used because of their potential for downregulating catabolism. Then, we monitored hACh production of healthy cartilage extracellular matrix proteins, total collagen (COL), and total glycosaminoglycan (GAG), as indicators of inflammation reduction. Our results show that simultaneous stretching and nutraceutical supplementation increased COL and GAG content by more than 11- and 21-fold, respectively.

Copyright © American Institute of Chemical Engineers. All rights reserved.

611296 Recapitulation of Bone Surface Tissue Complexity and Remodeling on Demineralized Bone Paper

Abstract:

It has been well documented that trabecular bone undergoes repeated surface remodeling under tight molecular regulation. However, how regulatory molecules form spatiotemporal profiles and direct bone remodeling cellular processes within trabecular bone cavities remains uncertain due to lack of relevant experimental models. Here, we report a new strategy to create tissue engineered in vitro bone tissue model that reproduces essential bone tissue complexity and physiological processes with high fidelity and control. An osteoid resembling demineralized bone paper supported osteoblasts to develop mineralized mature bone tissue and subsequently acquire a resting state bone lining cell phenotype. Vitamin D3 and prostaglandin E2 stimulation reverted lining cells to osteoblasts. Shifted regulatory molecular profiles induced differentiation of bone marrow mononuclear cells into osteoclasts. By layering stimulated osteoblasts above resting lining cells using demineralized bone paper insert with a spacer, we simulated a local activation of trabecular bone niche with defined spatiotemporal molecular gradients. Quantitative spatial mapping of bone remodeling activity substantiated that osteogenic cells forming extensive gap junctions are significantly less responsive to stimulative molecules than bone marrow mononuclear cells that function as individual cells. These findings deliver new insights into molecular and cellular regulation of trabecular bone remodeling, and the established model may enable to study various aspects of bone biology on endosteal niche.

Copyright © American Institute of Chemical Engineers. All rights reserved.

602351 Droplet Generation in 3D Printed Microfluidics Using a Hydrophobic Resin

Abstract:

Droplet microfluidics devices have proven useful as nanoliter-sized reactors, and in the area of bioengineering they serve as single-cell reactors for a variety of research applications. We are using nL-sized aqueous droplets suspended in oil as growth chambers for individual bacteria that are exposed to various antibiotics to assess the antibiotic susceptibility of the bacteria. We are also using novel 3D printing to create truly 3D fluidic channels for these nanodroplets, not simply a 2D-layout of channels in PDMS covered with a glass slide. Printing 3D microfluidic droplet devices allows for fluid channel flexibility and for rapid design iteration of devices using truly 3D geometries. However, many of the monomers used in this type of 3D printing are too hydrophilic to provide stability of the aqueous droplets, in that they often stick to the sides of the channel. Combined oil-water flows in these hydrophilic resins tend to form co-flowing streams where both phases wet the sides of the channel. For example, converging channels and pinch points that usually form distinct droplets in PDMS microchannels simply force both fluids against the walls without forming droplets. This study presents the development of a 3D printing resin monomer that when polymerized is sufficiently hydrophobic to produce stable water-in-oil droplet formation in our 3D printed microfluidic devices. We present the surface energies and mechanical properties of several resins investigated during this development. Examples of droplet generation will be shown as well as the integration of 3D geometry in the manipulation of aqueous droplet suspensions in 3D printed microfluidic devices.

Copyright © American Institute of Chemical Engineers. All rights reserved.

598493 Modeling Growth Kinetics of Cytotoxic T Lymphocytes for Cancer Immunotherapy in a High-Density Centrifugal Bioreactor System

Abstract:

Cancer was the second leading cause of death in the United States and worldwide in 2018. Traditional cancer treatments such as chemotherapy and radiation are designed to destroy cancer cells but often attack healthy tissue in the process. Alternative treatment methods include immunotherapy, which uses a patient's immune system to fight off cancers. In one form of cancer immunotherapy, known as adoptive cell therapy (ACT), immune cells called cytotoxic T lymphocytes (CTLs) are extracted, modified, rapidly expanded, and transferred into a patient. Current forms of ACT are expensive and inefficient, and there is a need for expansion methods that make ACT-based treatments widely accessible, scalable, and relatively affordable. To address this need, our research group has developed a high-density, lab-scale centrifugal bioreactor (CBR), which can rapidly expand infected CD8+ T cells from a bovine model. In other applications, the CBR has reached cell densities up to 2×10^8 cells per mL in an 11.4 mL chamber over the course of 7 days. In this study, we aimed to optimize CTL growth in the CBR by determining kinetic growth parameters based on the levels of glucose and inhibitory metabolites in the culture. It is hypothesized that if we can develop a reliable kinetic growth model from static culture, then it will be possible to predict optimal CTL expansion parameters for the bioreactor. Early kinetic studies were performed last fall in which six different glucose concentrations were tested, giving a maximum specific growth rate of $0.0112\ h^{-1}$ and a Monod constant of 2.12 mg glucose/dL. The experiments were repeated recently with different glucose concentrations and the results from those studies will be presented here. In conclusion, optimization of the kinetic growth models for the CBR will have a major impact on the availability and efficiency of patient-specific cancer immunotherapy.

Copyright © American Institute of Chemical Engineers. All rights reserved.

610202 Molecular Recognition and *In Vivo* Detection of Temozolomide and 5-Aminoimidazole-4-Carboxamide for Glioblastoma Using Fluorescent Nanosensors

Abstract:

The efficacy of chemotherapeutics such as temozolomide (TMZ) and its metabolic product 5-aminoimidazole-4-carboxamide (AIC) is often affected by the timing, quantity and frequency of dosages. There is strong interest in facilitating the ability to monitor efficacy in individual patients for specific subtypes of cancer. Real-time, dynamic measures of potency may also supplement or in some cases replace reliance on bio-imaging. Towards this end, in this work we develop new, synthetic molecular recognition sites for TMZ, and its decomposition product AIC, grafting them onto near infrared fluorescent nanoparticles capable of forming optode or other biosensor interfaces to monitor drug efficacy in real-time. Infrared fluorescent single-walled carbon nanotubes, wrapped using specific DNA oligonucleotides, are encapsulated within biocompatible, poly(ethylene glycol) diacrylate hydrogels and enable the selective recognition of an anti-cancer drug, TMZ, on U-87 MG human glioblastoma cells. In both solution phase and hydrogel form, the sensors exhibit a fluorescence response to TMZ with a detection limit of 30 M. Furthermore, U-87 cells exhibit no changes in viability for 7 days in contact with the hydrogel. The sensors were used to track the progression of glioblastoma death following TMZ administration. In addition, the sensors were able to monitor *in vivo* AIC and TMZ dynamics in SKH-1E mice. These results enable real-time detection of chemotherapeutic concentration and metabolism at the cellular and subcellular levels, with potential to increase the efficacy of cancer treatments.

Copyright © American Institute of Chemical Engineers. All rights reserved.

606459 An Enzymatic Electrochemical Biosensor for Real-Time Detection of Physiologically Relevant Nicotine Concentrations

Abstract:

The glucose biosensor, built upon the redox enzyme glucose oxidase, is the most commercially successful and studied enzymatic biosensor. However, the lack of available and functionally validated enzymes is prohibiting the development of redox-based sensors for other important analytes. Herein, we present the development and assessment of an electrochemical nicotine biosensor, using genomic screening to identify the gene for a known nicotine catabolizing redox enzyme. The resulting nicotine biosensor demonstrated a specific, sensitive, and stable operational profile with a limit of detection of 27 μM over the range of 0-200 μM. This range is well within the physiological concentrations of nicotine present in smoker urine. Specificity and cross-reactivity were measured against structurally similar compounds to nicotine as well as to known physiological by-products. Our results highlight that this novel enzymatic electrochemical nicotine biosensor possesses operational capabilities for monitoring of nicotine in physiologically relevant conditions. The screening methodology can be generalized for the discovery of enzymes for novel sensor development.

Copyright © American Institute of Chemical Engineers. All rights reserved.

608267 Simulations Guide Optimization of Electroenzymatic Biosensors for Neurotransmitters and Enable Proper Interpretation of Sensor Response *In Vivo*

Abstract:

Electroenzymatic biosensors for neurotransmitters are uniquely compact, sensitive, and selective devices that provide the neuroscience community with a method to monitor fluctuating chemical concentrations in the brain. Many studies primarily consider glutamate, the most common excitatory neurotransmitter, and study the relation between sub-second neurotransmitter dynamics and various stimuli, behaviors, or diseases. Challenges, however, have remained in the optimization of sensor design and in clarifying the significance of assumptions made during *in vivo* use. Accordingly, a number of simulations were carried out to guide the development of glutamate sensors and to test the limitations of the most prominent assumption, that sensor response can be linked to the immediate extracellular neurotransmitter concentration using a predetermined calibration factor. Recommended modifications to sensor design were developed and successfully implemented, improving sensitivity six-fold (up to 320 $nA/\mu M/cm^2$) with an accompanying order of magnitude decrease in response time (down to ~80 ms). Since even this improved response time is an order of magnitude slower than the timescale of synaptic neurotransmitter release, it is clearly wrong in many cases to assume that the use of a calibration factor to translate sensor response to neurotransmitter concentration without any signal deconvolution is accurate. To investigate the significance of this assumption and the conditions for which this practice remains reliable, the model was expanded with a new set of partial differential equations to describe the extracellular space of the brain. Considering the effects of variations in the local biological environment bordering a sensor, a range of representative parameters for some biological processes were selected. Comparisons were made between the theoretical responses expected from *in vitro* calibrations, tonic steady-state glutamate presence, and bolus-type glutamate releases that mimic synaptic release. Surprisingly, in steady-state cases, the sensor response could indicate a concentration over twice the actual value, while in cases where sensors were 10 μm or more from the source of glutamate being detected, sensor response was < 50% of the expected value. In some cases, transient concentrations due to a single synaptic-type release of neurotransmitters at a moderate distance from the sensor were not detected by the sensor at all, and when they were, the time course of signal decay did not match the rates of reuptake expected in the brain without significant signal deconvolution. Ultimately, results show that additional characterization of sensor performance *in vivo* may be necessary to enable proper interpretation of sensor data.

Copyright © American Institute of Chemical Engineers. All rights reserved.

611068 Theoretical and Experimental Studies on an Electrochemical Enzyme Immunosorbent Biosensor

Abstract:

Electrochemical immunosensors integrate biological recognition molecules (e.g., antibodies) with redox enzymes (e.g., horseradish peroxidase (HRP) to combine the advantages of immunoassays (extremely high sensitivity and selectivity) with those of electrochemical biosensors (reproducible, quantitative electrical output). Mechanistic mathematical models that describe the multiple mass-transfer and chemical reaction steps that give rise to the electrical output are needed to help design, optimize, and validate electrochemical immunosensors for medical applications. In this study, we developed such a model of an electrochemical immunosensor that used HRP as a reporter enzyme on the secondary and validated it experimentally.

The model consists of a system of non-linear mass-balance equations that describe simultaneous mass transfer and chemical reactions of the reactants (catechol and hydrogen peroxide) that HRP oxidizes, as well as the HRP product, O-quinone, which is reduced at the working electrode to give the amperometric output. A bi-substrate, ping-pong reaction mechanism was used to describe the enzymatic reaction, and the *Butler-Volmer* kinetic model was used to describe the reduction of o-quinone back to catechol at the working electrode's surface. The system of equations was solved numerically using the BVP4C function in MATLAB.

Experimental data for model validation was obtained using electrochemical immunosensors for a surrogate protein antigen (mouse IgG). The immunogens interface was fabricated on the gold working electrode of DropSens screen printed electrode arrays. The steady-state amperometric output of the resulting immunosensors was measured over a range of four independent variables: the concentrations of the two reactants, the voltage of the working electrode, and pH. The experimental results were used to determine optimal operating conditions and to validate the model. The validated model was then used to calculate Damkohler numbers and flux-control coefficients to identify mass-transfer and reaction steps that most strongly affected the magnitude of the immunosensors' output and the output's sensitivity to analyte concentration.

Copyright © American Institute of Chemical Engineers. All rights reserved.

610485 HTL Derived Biochar and Graphene Nanoplatelets for Biosensor Applications

Abstract:

HTL Derived Biochar and Graphene Nanoplatelets for Biosensor Application

Bharath K Maddipudi, Vinod Amar, Hope Dosch, Anuradha Shende and Rajesh V Shende*Department of Chemical and Biological Engineering, South Dakota School of Mines and Technology, Rapid City, South Dakota 57701, United States of America

***Corresponding Author:** Rajesh Shende, Department of Chemical and Biological Engineering, South Dakota School of Mines and Technology, Rapid City, South Dakota 57701, United States of America (E-mail: *Rajesh.Shende@sdsmt.edu*)

Abstract

Biosensors with Au electrode layer immobilized with peptides are currently being developed for the detection of microbial contamination. It is anticipated that the use of a low-*k* dielectric layer modified with graphene will improve the sensor sensitivity. The low-*k* dielectric layer can be modified with the biochar derived from hydrothermal liquefaction of different biomass substrates. In this investigation, HTL of lignocellulosic substrates was performed under non-catalytic and catalytic conditions and by-products were analyzed. The biochar was thermally treated in a tubular reactor under inert environment from 400-1100°C without any pretreatment. In other set of experiments, HTL derived biochar was acid treated and subsequently, thermal activation was performed. After the acid pretreatment and thermal treatment steps, the porosity, the specific surface area (SSA) and the morphology of the char was analyzed using BET and SEM/EDX. The char obtained after thermal treatment was combined with low-*k* dielectric layer and biosensor was fabricated. The sensor output was compared with the biosensor fabricated with low-*k* and graphene nanoplatelets. In addition, the sensor output was correlated with the char materials obtained under different thermal conditions. The results obtained on hydrothermal liquefaction of biomass substrates, characteristics of biochar before and after thermal treatment, biosensor fabrication and sensor output will be presented.

Copyright © American Institute of Chemical Engineers. All rights reserved.

604780 Construction of a Red Emission Bodipy-Based Probe for Tracing Lysosomalviscosity Changes in Culture Cells

Abstract:

A red emission BODIPY-based lysosome-targeted viscosity probe (Lys-VBOD) was synthesized for monitoring lysosomal viscosity in live cells. The increase of viscosity can restrict the twisting of double bond between indole
and BODIPY fluorophore, also restrict the rotating of single bond between quinoline and BODIPY, which lead to the enhancement of fluorescence intensity in high viscosity environment. The changes of fluorescence intensity
exhibit a good linear relationship with viscosity values. Besides, Lys-VBOD is quite stable in various pH solvents with given viscosity, suggesting that Lys-VBOD is suitable for applying in a wide pH range. Moreover, Lys-VBOD
can visualize the changes of lysosomal viscosity in live cells under the stimulus of dexamethasone. BSA titration experiment demonstrated that Lys-VBOD is stable in the presence of macromolecules and proteins. Notably, the
fluorescence lifetime of Lys-VBOD in large viscosity solvents is extremely high. These results indicate that LysVBOD is an excellent tool for detecing viscosity changes in vitro and live cells.

Copyright © American Institute of Chemical Engineers. All rights reserved.

610593 Super Sensitive Cerium Oxide-Based Composite Sensor for the Detection of Hydroxyl Radicals

Abstract:

Free radicals are recognized as essential molecules for upholding the function of normal cells. Immune systems, such as bacteria elimination and virus inactivation, are examples of the crucial role that free radicals play in human cells. Even though free radicals have benefits, an optimum production is key to maintain their advantages, as an excess can cause the development of oxidative stress conditions, such as cancer, Alzheimer's, and Parkinson's diseases. Among all free radical species, hydroxyl radicals (•OH) are the most reactive and dangerous, and they can be used as biomarkers for the detection of those diseases at their initial stages. Thus, a very sensitive analytical sensor is needed to detect and monitor low concentrations of •OH. The integration of an electrochemical technique with a sensing element is regarded as a promising method for •OH detection due to a rapid and direct measurement without the pretreatment of samples. Thus, in this study, a glassy carbon electrode was modified with a sensing element comprised of ultrasmall cerium oxide nanoclusters (<2 nm) grafted onto a superconductive carbon (Figure 1a). XPS results showed 46.3% Ce^{3+} sites, which are those able to react with •OH. Cyclic voltammetry (CV) was implemented to characterize the interaction of the composite sensor with •OH, and they showed that the composite sensor was able to differentiate between solutions with and without •OH by an increased redox response in presence of •OH (Figure 1b). Furthermore, we observed a linear relationship between the redox response and the concentration of •OH in the range of 0.1 mM to 10 mM, with a limit of detection as low as 0.1 μM when using only 0.5 wt% CeO_x loading.[1]

(1) A.C. Alba-Rubio, D.S. Kim, S. Duanghathaipornsuk; Super sensitive sensor for the detection of hydroxyl free radicals with scavenging properties. Application number 62/977,945 (2020).

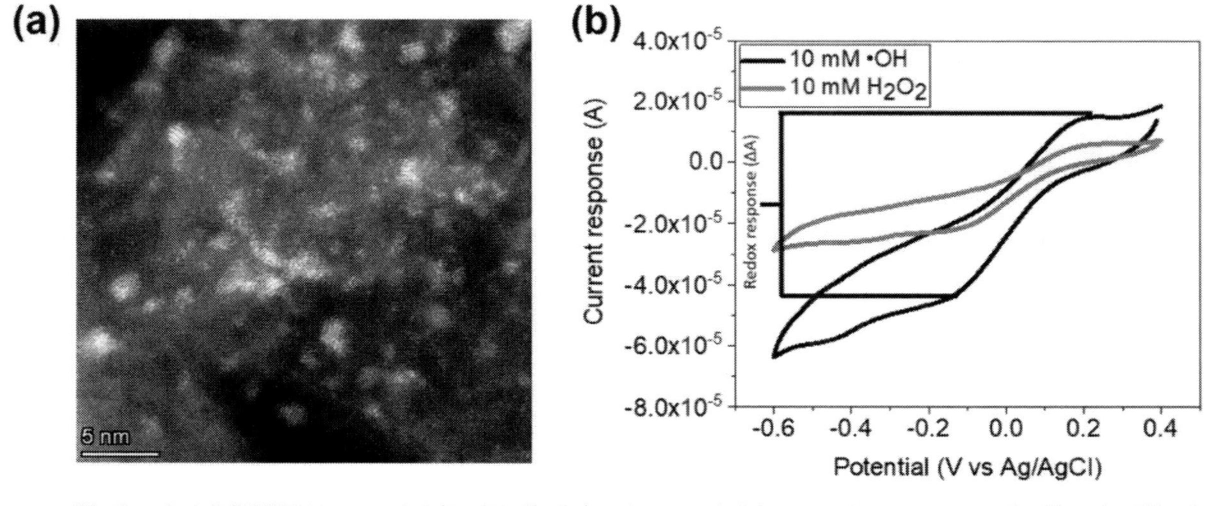

Figure 1. (a) STEM image of 0.5 wt% CeO_x/Carbon and (b) Current response to hydroxyl radicals.

Copyright © American Institute of Chemical Engineers. All rights reserved.

611380 High Throughput Antibiotic Susceptibility Testing with Optical Nanosensors

Abstract:

Infections in a clinical setting can result from biofilm-forming species, which presents two key challenges: increased antimicrobial resistance, and poor diagnostic approaches to categorize antimicrobial susceptibility. Traditional antimicrobial susceptibility assay approaches are built upon planktonic bacteria (bacteria that are growing in solution). Biofilms are bacteria that grow in a community on a surface, and this community structure changes resistance to antimicrobials that is not reflected in these planktonic assays. While current biofilm-based assays exist, they are both time and material intensive, and can still provide antibiotic susceptibility results that may not reflect the actual biofilm response. Here we present our nanosensor based approach to monitor biofilm metabolic response during the administration of antibiotics. We grow biofilms with oxygen-monitoring nanosensors embedded throughout the film, and these nanosensors continuously report out the metabolic state of the biofilm. As we add antibiotics, we can measure what concentration can stop metabolism, and infer the minimum biofilm inhibitory concentration (MBIC) of the specific strain. Our diagnostic approach functions in 96 well plates with minimal materials and time (important for high throughput screening), and works with both lab and clinically relevant strains (demonstrated with Pseudomonas aeruginosa).

Copyright © American Institute of Chemical Engineers. All rights reserved.

609958 Multi-Modal Single-Cell Analysis As a Basis for Precision Therapy Targeting Intratumoral Heterogeneity in Glioblastoma

Abstract:

Glioblastoma (GBM), the most aggressive and lethal form of primary brain cancer in adults, is a prominent example of how a heterogeneous tumor cell population hinders our ability to treat cancer[1–4]. An unfortunate consequence of this intratumoral heterogeneity is the failure to improve survival rates using precision medicine approaches that rely on bulk-level DNA sequencing. Bulk-level profiling methods fail to capture the heterogeneous composition of a tumor. However, advances in single-cell level genomic profiling technologies (e.g. single-cell ATAC-seq, single-cell RNA-seq) and integration of such data enable the multi-modal characterization of 'cell state', i.e., the metastable phenotype of a cell at a given moment. In GBM, tumor cells evolve during therapeutic intervention, possibly into drug-resistant states, along trajectories that are as yet unknown. Towards understanding these cell-state changes and identifying novel targets to eliminate the various tumor cell subpopulations, we created a patient-derived xenograft (PDX) mouse model of an individual patient's GBM tumor biopsy to simulate patient-derived tumor cell response to chemotherapy (temozolomide) and radiotherapy, which are part of the standard of care. We characterized tumor cell response using single-cell ATAC-seq and RNA-seq analysis and compared the cell states of xenograft samples 1) prior to treatment, 2) 24hrs post-treatment, and 3) 72hrs post-treatment to those of the corresponding parental tumor. Moreover, we applied the Systems Genetics Network AnaLysis (SYGNAL) platform[4] to identify distinct epigenetic programs that distinguish phenotypically diverse tumor samples across conditions. Further, we identified several drugs that target specific regulators associated with these epigenetic programs, which we will be testing in corresponding tissue-culture slices for validation. This proof-of-concept work will provide the basis for our envisioned goal: the development of a modeling and analytical system that enables single-cell characterization of an individual patient's tumor and its response to drugs. Ultimately, the information gathered from this systematic analysis of an individual tumor would inform clinical treatment and enable the ultimate goal of precision medicine.

References

1. Patel, A. P. *et al.* Single-cell RNA-seq highlights intratumoral heterogeneity in primary glioblastoma. *Science (80-.).* **344**, 1396–1401 (2014).
2. Weathers, S. P. & Gilbert, M. R. Current challenges in designing GBM trials for immunotherapy. *J. Neurooncol.* **123**, 331–337 (2015).
3. Ellis, H. P. *et al.* Current Challenges in Glioblastoma: Intratumour Heterogeneity, Residual Disease, and Models to Predict Disease Recurrence. *Front. Oncol.* **5**, 251 (2015).
4. Plaisier, C. L. *et al.* Causal Mechanistic Regulatory Network for Glioblastoma Deciphered Using Systems Genetics Network Analysis. *Cell Syst.* **3**, 172–186 (2016).

Copyright © American Institute of Chemical Engineers. All rights reserved.

601531 Modeling and Manufacturing Methods for Downsizing a Centrifugal Bioreactor for Use in Cancer Immunotherapy

Abstract:

In the fight against cancer, it is of major interest to shift treatments to targeted, systemic therapies such as immunotherapy, in which we harness the power of our own immune systems. With cell-based therapies, though, rapid expansion of cell cultures that maintain phenotypic activity is necessary to achieve the high cell requirements per treatment. Over the years, our group has produced a centrifugal bioreactor (CBR) system that can be used for a number of applications including suspension and anchorage-dependent cell culture, but we are particularly interested in using it to manufacture cytotoxic T lymphocytes (CTLs) for cancer immunotherapy. It is a modified COBE Spectra (now Terumo BCT) blood apheresis system, where the centrifugal force and fluid forces are balanced to act similarly to a fluidized bed and maintain high cell densities in the reactor chamber. In addition to the forces acting within the reactor chamber, the CBR's ability to produce high-density cultures is also due to operating in perfusion mode, which allows the cells to be continuously refreshed with medium and nutrients. In previous work, the CBR has been used to rapidly expand cultures up to 2×10^8 cells/mL in an 11.4 mL chamber over the course of 7 days. Currently, the CBR is at a larger scale than ideal, as half of the system exists outside of a biosafety cabinet, and a major issue in using the CBR for T cell expansion is the numerous sources for potential contamination. While the current system does work with careful technique, we have shifted our recent efforts to miniaturizing the bioreactor to fit within a biosafety cabinet. Various combinations of the cell chamber volume, rotor disc radius, centrifugal force, and feed flow rate can be modeled to determine optimal parameters for maintaining fluidization. In this presentation, we will share mathematical modeling for downscaling this centrifuge-based bioreactor and manufacturing methods for the prototype system.

Copyright © American Institute of Chemical Engineers. All rights reserved.

607954 TGF-β Inhibition Combined with Cytotoxic Nanomedicine Normalizes Triple Negative Breast Cancer Microenvironment Towards Anti-Tumor Immunity

Abstract:

Normalizing the tumor microenvironment (TME) consists of restructuring immature and compressed vessels and in the process, improving their functionality. These normalized vessels have increased perfusion and increased drug delivery as well. The techniques have been shown to be successful at TME normalization, and we believe a combination of nanomedicine and immunotherapy treatments would have even stronger results that enhance the efficacy of immunotherapy. So, we chose tranilast, the TGF-β inhibitor, which is an approved antifibrotic and antihistamine drug to combine with Doxil nanomedicine and doxorubicin for two triple-negative breast cancer mouse models. After observing the normalization effects of the combinatorial treatments, we also analyzed the effects on the immune checkpoint inhibitors. The combination treatment of tranilast and Doxil resulted in a decrease of mechanical forces, which we observed by quantifying the area fraction of components of the extracellular matrix. Additionally, we detected an increase in pericyte coverage, blood vessel diameter, and T cell migration from the treatment. These factors indicate TME normalization and resulted in a decrease in tumor growth. Specifically, the T cells migrated far from immunosuppressive CAFs, which caused a greater T cell distribution that can increase the efficacy of immunotherapies. Finally, the combinatorial treatment increased the ratio of anti-tumor M1-like macrophages to immunosuppressive M2-like macrophages. A triple combination of tranilast, Doxil, and a cocktail of immune checkpoint blocking antibodies resulted in the most reduced tumor growth, displaying the positive effects of TME normalization strategies on immunostimulation.

Copyright © American Institute of Chemical Engineers. All rights reserved.

605162 Unveiling the Cell Cycle Mediated Apoptosis Mechanism and Therapeutic Efficacy in Breast Cancer Cells

Abstract:

HER2-positive breast cancer accounts for almost 15-20% of the all the breast cancer patients. This subgroup of patient's tumor is characterized by HER2/ERBB2 gene copy number amplification resulting in HER2 overexpression. HER2 activates RAS-RAF-MAPK and PI3K-AKT mitogenic signaling pathways. This also leads to the deregulation of cell cycle regulatory genes including CCND1, CDK4/6, CCNE1, loss of tumor suppressor gene PTEN and loss of p53 (Figure 1). The molecular and genetic aberrations are heterogeneous and often acquire resistance to the intervened therapy in early stage. It is hypothesized that the mechanism of resistance is regulated by the cell cycle. HER2-positive breast cancer cells were arrested in different phases of the cell cycle G_0, G1, S and G2-M using serum deprivation, Trastuzumab or Rapamycin, 5-fluorouracil and Paclitaxel or Doxorubicin, respectively. We also used Trastuzumab functionalized Paclitaxel drug nanorods (PTXNR-TTZ) that arrest cells in G2-M phase and induce apoptosis synergistically on HER2-positive breast cancer cells. Untreated cells were mostly arrested in G_0/G1 phase. We investigated the genetic and molecular alterations in cells using Next Generation Sequencing (NGS) technique. Our study includes the discovery of key driver genes and genetic network responsible for cell cycle dependent therapeutic resistance and therapeutic outcomes.

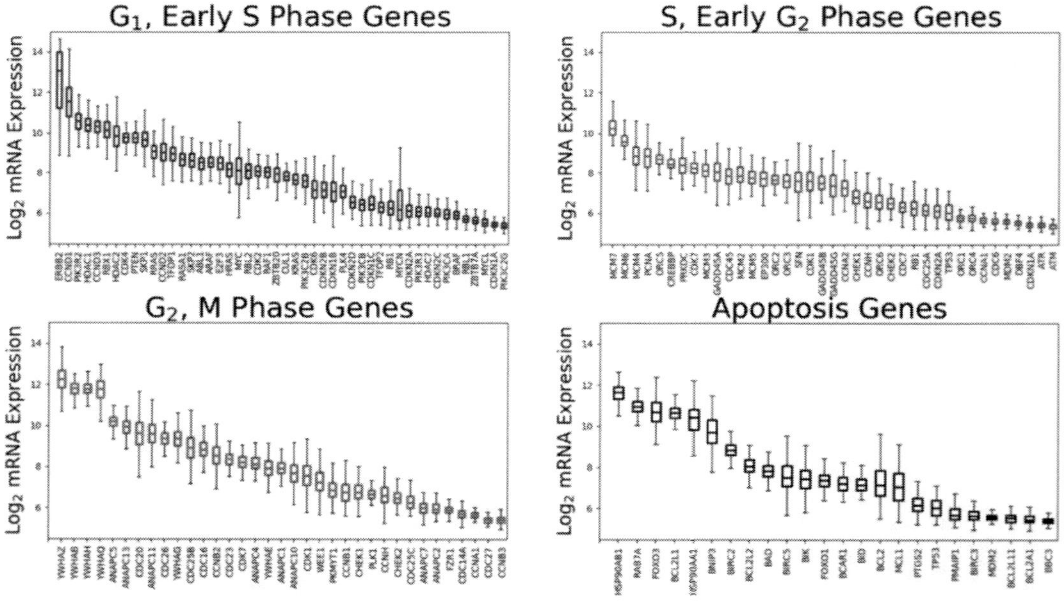

Figure 1. Genetic aberration in cell cycle and apoptosis related genes can be utilized to predict the potential therapeutic target in HER2-positive breast cancer. The patient's data (n=220) is collected from the METABRIC HER2-enriched patient dataset.

Copyright © American Institute of Chemical Engineers. All rights reserved.

602379 Measurement of Cancer Cell Drug Response with Quantitative Phase Imaging

Abstract:

There are no biomarkers that can accurately predict chemotherapy response in advanced cancer patients and less than 10% of patients with a detected targetable mutation are eligible for a clinical trial. There is a need for new diagnostic methods that can accurately stratify high-risk patients to effective, FDA-approved therapies. Our current patient-derived models for assessing tumor drug response involve expanding patient tumor cells as 3D patient derived organoids (PDO) in Matrigel or using *in vivo* drug sensitivity studies with patient-derived xenograft models (PDX). These experimental models typically exhibit the same phenotype and molecular alterations *in vivo* and *ex vivo* and have the same drug responses as in the patient. However, these methods require 1-8 months to obtain drug sensitivity profiles making this impractical for patient care. We have tested a functional assay based on quantitative phase imaging (QPI) with the capability to predict cancer cell response to therapy within 24 h of data collection. QPI is a label-free microscopy method that uses light to measure the mass of single cancer cells. By tracking the mass of growing cancer cells over time, QPI provides a rapid measurement of cell growth or death that can be used to monitor response/resistance to different drug therapies. QPI also provides data on heterogeneity of response that may indicate emergence of resistant subpopulations. We have developed and validated a QPI imaging system for dedicated screening of breast cancer cell response to therapy. We validated this approach against gold standard measurements and will show response profiles to a panel of FDA-approved chemotherapies. The overall goal of our work is to provide real-time feedback to oncologists on drug sensitivity/resistance and resistant subpopulations.

Copyright © American Institute of Chemical Engineers. All rights reserved.

610858 Synergistic Combinations Using APE1 Inhibitors for Ablation of Advanced Melanoma

Abstract:

Cancer is a debilitating disease which is diagnosed in nearly 1.7 million Americans each year. Nearly 200,000 of these are attributed to bladder(4th common), skin (5th common) and ovarian Cancer (7th common). These cancers result in the death of 40,000 people annually in the United states of America. A popular strategy to overcome these cancers has been the use of small molecules to inhibit the molecular chaperone heat shock protein 90(HSP90), which protects cancer cells from multiple forms of environmental stress through its ability to regulate a multitude of signaling and growth pathways. Unfortunately, the in-vitro efficacy of targeting HSP90 has not translated well into the clinic with the most promising results shown in breast and lung cancer. This is mostly attributed to resistance mechanisms developed due to expression of proteins like HSP70, HSP27 and HSF-1. Here we show a novel combination of inhibitors of HSP90 and APE1, A DNA repair/redox function protein which led to synergistic cell death in several cancer cells both using 2d proliferating and 3D dormant cell cuture systems. The combination treatment induced a significant drop in mitochondrial membrane potential and significant increase in Reactive oxygen species(ROS) generation respectively. Based on our hypothesized mechanism we measured change in expression of proteins like PTEN, NF-ΚB, AP-1 etc post combination treatment and were able to observe a significant change in expression levels. The Combination treatment demonstrated significant reduction in growth of aggressive human melanoma tumors, containing both BRAF V600E and NRAS Q61K mutations, in NOD SCID gamma mice compared to single agent treatments acting alone. The treatment also significantly abrogated metastasis of cancer cells to the lungs and liver compared to individual treatments acting alone. Our results indicate that the synergistic inhibition of APE1 and HSP90 proteins is a promising combination treatment for ablation of aggressive tumors.

Copyright © American Institute of Chemical Engineers. All rights reserved.

607552 Efficient and Safe Delivery of Cancer Chemotherapeutics By DNA Fragments

Abstract:

Small molecule pharmaceutical treatment of cancer has several issues, including lack of bioavailability, insolubility, and systemic toxicity. While nano-scale drug delivery systems have risen to the challenge in overcoming issues faced by free small molecules by increasing the retention time of drugs and decreasing toxicity; issues of immunogenicity, non-biodegradability, and toxicity still remain due to the materials used in their manufacture. Furthermore, rarely are these nanoparticles produced without the incorporation of several intensive processes that end up making them expensive to synthesize and difficult to scale up for clinical or commercial use. In addition, disappointingly low therapeutic payload is a common concern. Thus, we looked towards natural biopolymers and wondered if it was possible to simply use nucleic acids themselves as a drug delivery vehicle for cancer chemotherapeutic delivery. DNA fragments are readily produced from a natural source and offer the desired properties for the delivery of small molecule therapeutics. A model cancer chemotherapeutic was complexed with DNA fragments in a scalable and cost-effective manner. The resulting nano-scale complexes demonstrated improved anti-cancer efficacy with substantially lowered toxicity *in vitro* and *in vivo*, in comparison with free drug and a commercially available liposomal formulation. This talk also highlights the characterization of the complexes at a molecular level and a mechanistic study on the complexes' interactions with cancer cells *in vitro* and *in vivo*. The results of this study encourage an elegant and simple approach to drug formulation that utilizes the natural properties of nucleic acids to address the fundamental challenges in cancer chemotherapy. Finally, our new formulation may offer immediate implications for rapid clinical translation and commercialization.

Copyright © American Institute of Chemical Engineers. All rights reserved.

605681 Programming Structure-Function Relationships in Protein Megamolecules for Prodrug Cancer Therapy

Abstract:

Antibody therapeutics have revolutionized the way we treat cancer due to their ability to specifically target tumors, leaving healthy cells unaffected. However, it is difficult to chemically modify the size, shape, valency, and structure of antibodies due to a lack of available tools. This limitation in programmability contributes to factors such as immunogenic responses, inability to penetrate solid-tumors, and high dosage requirements. Our lab has developed a protein assembly method known as the 'megamolecule' approach that can address this limitation. The megamolecule strategy provides precise, programmable control over therapeutic structures through enzyme mechanism-based covalent inhibition of fusion proteins containing a connector protein domain (cutinase or SnapTag) and an active protein, such as an antibody fragment. We used the megamolecule approach to create nanobody-enzyme conjugates for prodrug cancer therapy with precise control over both antibody and enzyme stoichiometry. As a first system, we used an anti-HER2 nanobody and an enzyme, yeast cytosine deaminase, that converts a nontoxic prodrug into a chemotherapeutic. We evaluated the colloidal behavior, antibody binding kinetics, enzyme activities for prodrug conversion, and cell-specific efficacy of the megamolecules in both adherent and spheroid culture. We found that increasing the number of enzyme domains within a molecule allowed for intramolecular dimerization of the yeast cytosine deaminase domains, resulting in a 7.5-fold improvement in the catalytic efficiency of the two-enzyme bearing protein therapeutic, as confirmed by enzyme assays and transmission electron microscopy. This improvement was followed by an increase in prodrug activation, resulting in a 4-fold cytotoxic improvement in adherent cell culture. Further, we generated fluorescent megamolecules with variable avidity to visually track and calculate the binding affinity of multi-domain nanobody megamolecules to HER2+ cancer cells in live cell culture using confocal microscopy. Our findings confirmed cell-specific binding and cytotoxicity of HER2+ cells compared to HER2- cells. We found that the addition of a second binding domain within the molecule resulted in a 5-fold reduction in the dissociation constant of the therapeutic for HER2. Studying the binding kinetics and solid tumor penetration of protein scaffolds using live-cell confocal microscopy has the potential to yield the design rules of cell targeting protein therapeutics. By enabling control over the specificity, orientation, and stoichiometry of active domains within a protein nanostructure, megamolecules represent next generation cancer therapies whose structure can be tailored for optimum activity.

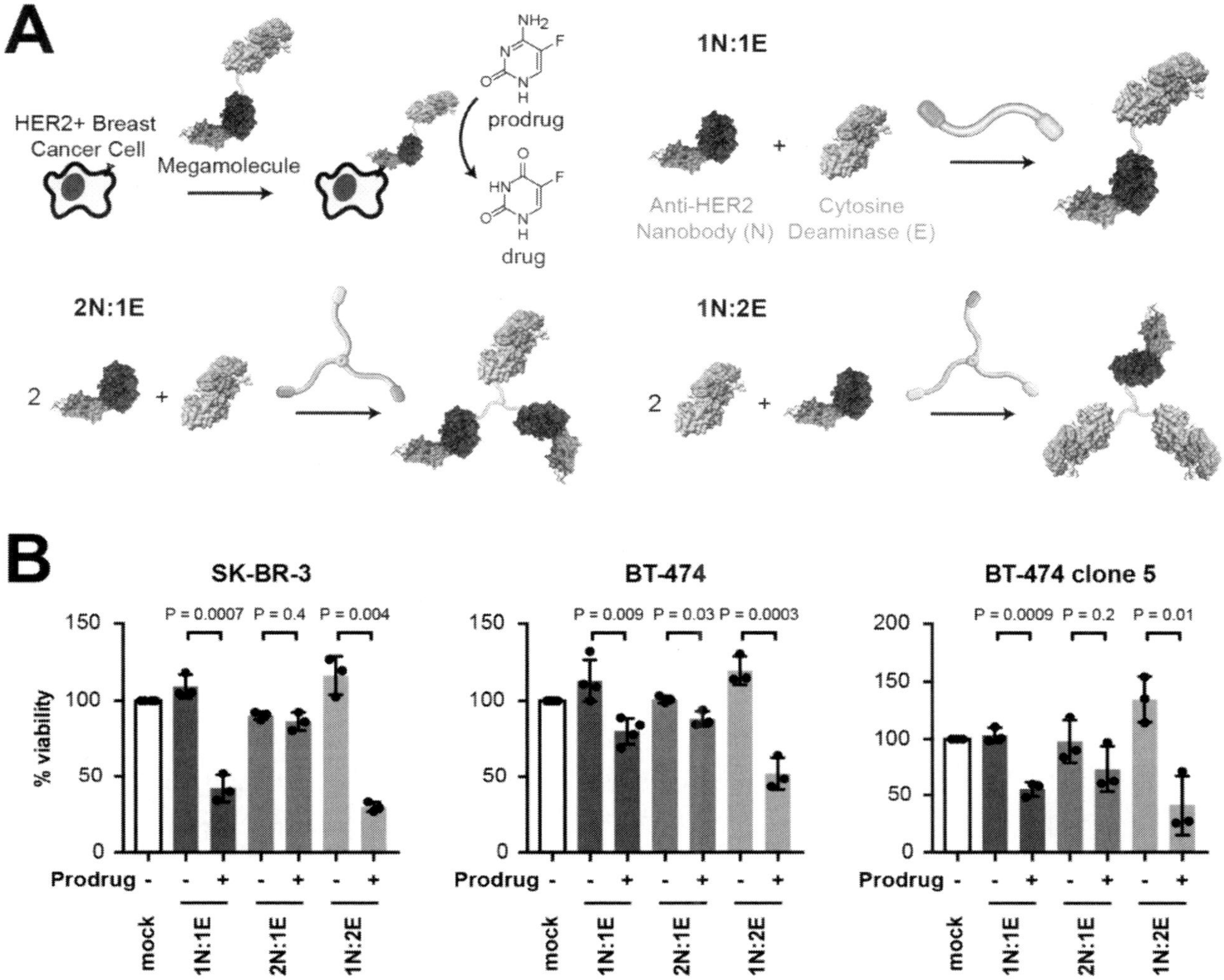

Copyright © American Institute of Chemical Engineers. All rights reserved.

611203 Point-of-Care Low-Cost Cancer Therapeutic Production and Monitoring Efficacy in Human Serum with Cell-Free Protein Synthesis: Toward on-Demand Treatment of ALL

Abstract:

Acute lymphocytic leukemia (ALL) is the most common form childhood cancer. Fortunately, administration of bacterial asparaginase (ASNase) has greatly improved survival rates, however these therapeutics have high incidence of immunological neutralization and serum activity must be monitored for most effective treatment regimens. Here, improvement in cell-free protein synthesis (CFPS) of FDA approved l-asparaginase (crisantaspase) is reported by employing an aspartate-fed-batch reactor format. A CFPS-based ASNase activity assay as a tool for therapeutic regimentation and production quality control is also presented. This work suggests that shelf-stable and low-cost *Escherichia coli*-based CFPS reactions may be employed on-demand to 1) synthesize biologics on-site for patient administration, 2) verify biologic activity for dosage calculations, and 3) monitor therapeutic activity in human serum during the treatment regimen. This work demonstrates the potential of CFPS as both a production platform and biosensor for treatment and treatment monitoring in the clinical setting.

Reference: Hunt JP, Wilding KM, Barnett RJ, Robinson H, Soltani M, Cho JE, Bundy BC. Engineering Cell-Free Protein Synthesis for High-Yield Production and Human Serum Activity Assessment of Asparaginase: Toward On-Demand Treatment of Acute Lymphoblastic Leukemia. Biotechnol J. 2020. 15(4):e1900294.

Copyright © American Institute of Chemical Engineers. All rights reserved.

606167 Deposition of Oxide Coatings Using Non-Thermal Atmospheric Pressure Plasmas

Abstract:

Atmospheric pressure plasmas (APPs) have shown great promise in a variety of manufacturing applications. APPs offer unique opportunities for the industrial application of plasmas, due to the versatility of the technique, the ease of scalability, and the opportunity to perform the technique point of manufacturing. This work focuses on depositing zirconia-silica bilayer coatings as a replacement for hexavalent chromium-based coatings. A new non-thermal process to deposit zirconia was developed using APPs enabling processing on temperature sensitive substrates. Electrochemical corrosion performance will be compared to other passivation and barrier coatings. Additionally, we investigate APPs to enable lightweight manufacturing for the automotive industry by depositing a thin film of silica that acts as a water intrusion barrier on bonds joints between dissimilar materials. The adhesion tests performed after water soak meets the industry benchmarks. These processes aim to replace wet chemical processes often used by the automotive and aviation industries. The mechanical properties of the films are tested using lap shear, both before and after a water soak test, which successfully demonstrates the films capabilities as a water barrier. The chemical composition is confirmed using both XPS and RBS, and the films morphology is examined using cross sectional SEM.

Copyright © American Institute of Chemical Engineers. All rights reserved.

608084 Treatment and Extraction of Copper from Electronic Waste Via Induced Morphological Changes Utilizing Supercritical CO_2

Abstract:

The management of electronic waste (e-waste) has become a significant challenge faced by humanity. Globally, the number of discarded computers, phones, and other appliances, often defined as Waste Electrical and Electronic Equipment (WEEE), doubled between 2009 and 2014 to 41.8 million tons per year. Currently, e-waste disposal factories tend to recycle the non-toxic, valuable components of e-waste (metals) and release the toxic components into the surroundings via pyrometallurgy (burning). Chemical treatment, known as hydrometallurgy, is also used and involves large amounts of corrosive acids. As the amount of e-waste grows, the availability of easily mineable elements, particularly metals, decreases, disturbing the overall material cycle. Thus, new, environmentally-friendly approaches to recover valuable components from e-waste are urgently needed. This study focuses on the use of supercritical CO_2 ($scCO_2$) and sulfuric acid, 2 M H_2SO_4, as a novel solvent system for the treatment of e-waste, specifically in the extraction of copper. Printed circuit board (PCB) was used as the e-waste of study and is generally composed of 40% metals, 30% plastics, and 30% refractory material. Through the synthesis of metal and polymer composites, which reduced the complexity of PCB, the effects of $scCO_2$ in the solvent system were investigated. The results showed that $scCO_2$ and acid induced a permanent morphological change in the polymer component, creating voids and causing delamination. This finding was translated to the actual waste PCB system and a novel two-step process was developed. This two-step process involved the pre-treatment of small pieces of waste PCB with $scCO_2$ and 2 M sulfuric acid at 120 °C and 148 atm for 30 minutes followed by leaching of the pre-treated PCB in a solvent containing 2 M H_2SO_4 and 0.2 M H_2O_2 at ambient conditions.Experimental results showed that 82% of the copper contained in the PCB was extracted in under 4 hours. The characterization of the PCB demonstrated that the pre-treatment with $scCO_2$ and acid induced the crystallization of the plastics (polymer component), creating pores and weakening the structure of the PCB, thus enhancing transport of the solvent to the buried metal interfaces. This novel process utilizing $scCO_2$ minimizes intensive physical processing (e.g. grinding of the PCB) and reduces acid usage during the extraction of Cu from e-waste, presenting a greener alternative to the current methods of recycling of metals from e-waste, which tend to be energy-intensive and hazardous.

Copyright © American Institute of Chemical Engineers. All rights reserved.

596648 Annealing Carbon By Pulsed Laser Light

Abstract:

Laser processing of materials is not new, but, using lasers to anneal, bond and otherwise transform carbons is. Fundamental understanding of the dependence upon carbon structure, morphology and chemistry is critical to implementing this technology into manufacturing and processing applications. In this work a Q-switched Nd:YAG laser and a continuous wave CO_2 laser are used to anneal carbon materials. Lasers provide rapid heating and cooling with high temporal control. The extent of transformation is kinetically controlled by time above the threshold temperature for transformation. Enabling the production of carbon structures not possible via traditional furnace annealing. The potential technological importance of laser annealing carbon is demonstrated. Examples of material processing and synthesis not possible via traditional furnace annealing are provided. To resolve the nanostructure changes HRTEM is employed.

Perhaps the most over looked application of laser annealing carbon is the ability to do so continuously and rapidly. Graphitization furnaces operate as a batch processing system. Time from start to completion is on the order of a day due to the slow (~25 °C/min) heating rates, long hold times, and slow cooling of the heavily insulated furnace. Additionally, these furnaces require routine maintenance and replacement of expensive specialty graphite heating elements. CO_2 laser annealing provides equivalent material transformation on the order of seconds and modern CO_2 lasers are designed for years of maintenance free use [1]. Although laser annealing is limited to thinner material due to limited heating depth, materials can be annealed continuously and with potentially high throughput.

The material was heated to 2,600 °C in 1.4 ms Under the action CO_2 laser radiation carbon can be heated to 2,600 °C in 1.4 ms. After 20 s, the structure is equivalent to that obtained from furnace annealing at 2,600 °C for 1 hr. as has been shown elsewhere [2]. In contrast a Q-switched Nd:YAG laser with pulse width ~ 8 ns heats carbon materials to graphitization temperatures with a heating rate on the order of a few 10^{11} °C/s. The peak temperature is controlled by the laser pulse energy; typical temperatures range between 2,500 °C to the C_2 sublimation temperature of 4,184 °C [2,3]. The time at elevated temperature is limited (time above 2,000 °C is 1.5 μs). On this time scale, the long-range material motions are kinetically restricted. Consequently the Nd:YAG laser annealing trajectory deviates from traditional furnace pathways. The limited time at elevated temperature can be used for the purpose of surface modification. Surface modification via kinetically limited oxidation is one potential application. Another application is making connections between carbons without additional material. Laser sintering a mixture of carbon materials could result in a wide range of potential applications.

Heat-Treatment-Temperature (HTT) and resulting carbon structural transformation has been extensively studied. A detailed quantification of HTT and resulting material annealing was provided by Oberlin in 1984 [4]. The four stages outlined in Oberlin's HTT diagram (temperature dependent) are believed to be separated by "very rapid" transitions [8]. So rapid, in fact, that intermediate structures are not observable on furnace heating timescales (several minutes minimum) due to slow ramp rates and have thus been ignored. These intermediate structures remain unexplored and represent an entirely new class of carbon materials. It has been demonstrated that CO_2 laser heating results in equivalent end structure as compared to furnace annealing [1]. The high temporal control of

CO_2 laser annealing allows for exploration of such intermediate structures that can be captured by the rapid heating and cooling timescales. In summary lasers are poised to be instrumental in the advancement of carbon science and technology.

References

[1] J.P. Abrahamson, Pulsed laser annealing of carbon, Ph.D. Thesis, Penn State, 2017.

[2] J.P. Abrahamson, M. Singh, J.P. Mathews, R.L. Vander Wal, Pulsed laser annealing of carbon black, Carbon 124 (2017) 380–390. doi:10.1016/j.carbon.2017.08.080.

[3] H.A. Michelsen, C. Schulz, G.J. Smallwood, S. Will, Laser-induced incandescence: Particulate diagnostics for combustion, atmospheric, and industrial applications, Prog. Energy Combust. Sci. 51 (2015) 2–48. doi:10.1016/j.pecs.2015.07.001.

[4] A. Oberlin, Carbonization and graphitization, Carbon 22 (1984) 521–541. doi:10.1016/0008-6223(84)90086-1.

Copyright © American Institute of Chemical Engineers. All rights reserved.

604429 An All-in-One Approach for Training Deep Learning-Based Control Laws

Abstract:

Model predictive control (MPC) provides a general framework for controlling nonlinear systems. However, the high online computational latency limits the scope of its application for systems with fast dynamics. One way to enable the real-time application of MPC is to approximate the implicit MPC control law with a neural network [1,2,3]. The hypothesis of this approach is that a deep learning neural network can sufficiently approximate the nonlinear behavior of MPC, but at a negligible online computational cost that is significantly better than that of a full-fledged MPC. For linear systems with quadratic cost function, a neural network with rectified linear units as activation functions can exactly represent piecewise affine optimal control laws [4].

An essential aspect of the above-mentioned approaches is the training of large dimensional neural networks. Many researchers follow the "optimize then train" protocol. The idea in these algorithms is to simulate the optimal controller, generate corresponding state and optimal control action data pairs and then use these data to train a neural network via supervised learning to approximate the original optimal controller. Unfortunately, this conventional "optimize then train" protocol is not guaranteed to work on nonlinear systems. For nonlinear systems, it is possible that multiple optimal control actions exit for the same initial state. In this case, the MPC control law is a set-valued function. The non-uniqueness of the state-action data pairs makes it impossible to learn the underlying MPC control law through this two-step approach, as illustrated in Figure 1. Even if the uniqueness of optimal control actions is guaranteed, the optimal control problem might have multiple local optimal control actions, which are typically identified using local solvers. Moreover, for general nonlinear systems, no neural network structure with finite neurons can guarantee an exact representation of the MPC control laws. The training error of the neural network can lead to sub-optimal control actions and state constraint violations. The resulting errors could potentially accumulate over time and eventually making the neural network deviate significantly from the optimal control.

We propose a new "optimize and train" method that combines the steps of data generation and neural network training into one single large-scale stochastic optimization problem. This approach directly optimizes the closed loop performance of the DNN controller over a number of possible initial states. Directly optimizing the control law instead of control actions avoids the difficulty caused by the nonuniqueness of optimal control actions, as shown in Figure 2. Moreover, all constraints are guaranteed to be satisfied for all initial state scenarios considered in the optimization, while this guarantee does not hold for the conventional "optimize then train" method. We also propose methods to ensure constraints satisfaction for all possible initial states. The critical challenge is the solution a large-scale nonlinear stochastic optimization problem. However, tremendous progress has been made recently in the field of stochastic optimization that allows us to design and implement efficient local and global algorithms [5,6]. Besides directly using stochastic optimization solvers, we also proposed a recurrent neural network formulation for nonlinear systems with input constraints and soft state constraints. This approach can take advantage of existing packages (e.g., TensorFlow) in the machine learning community and can potentially tackle extreme-scale problems efficiently with GPUs. The benefits of our all-in-one approach over the conventional ``optimize then train" protocol is illustrated through numerical results.

References:

[1] Parisini, T., & Zoppoli, R. (1995). A receding-horizon regulator for nonlinear systems and a neural approximation. Automatica, 31(10), 1443-1451.

[2] Csekő, L. H., Kvasnica, M., & Lantos, B. (2015). Explicit MPC-based RBF neural network controller design with discrete-time actual Kalman filter for semiactive suspension. IEEE Transactions on Control Systems Technology, 23(5), 1736-1753.

[3] Chen, Y., Shi, Y., & Zhang, B. (2018). Optimal control via neural networks: A convex approach. arXiv preprint arXiv:1805.11835.

[4]Karg, B., & Lucia, S. (2018). Efficient representation and approximation of model predictive control laws via deep learning. arXiv preprint arXiv:1806.10644.

[5] Cao, Y., Zavala, V. M., & D'Amato, F. (2018). Using stochastic programming and statistical extrapolation to mitigate long-term extreme loads in wind turbines. Applied energy, 230, 1230-1241.

[6] Cao, Y., & Zavala, V. M. (2019). A scalable global optimization algorithm for stochastic nonlinear programs. Journal of Global Optimization, 75(2), 393-416.

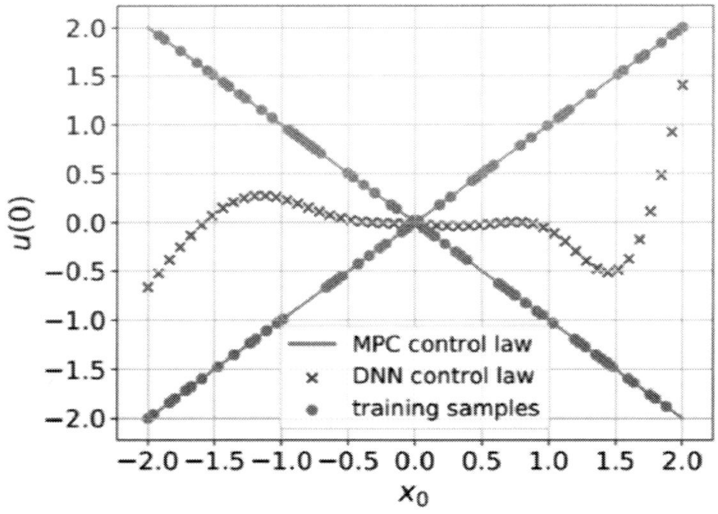

Fig. 1. DNN control law trained with "optimize then train" method

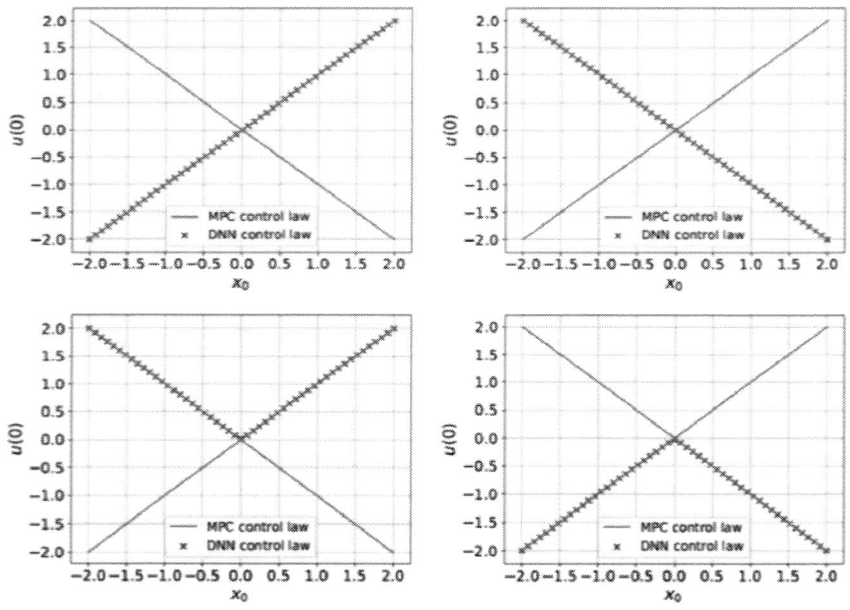

Fig. 2. DNN control law trained with "optimize and train" method

Copyright © American Institute of Chemical Engineers. All rights reserved.

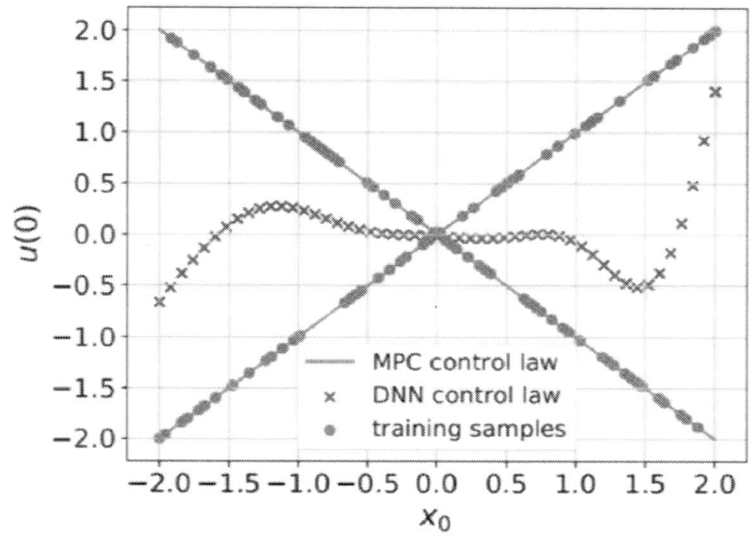

Fig. 1. DNN control law trained with "optimize then train" method

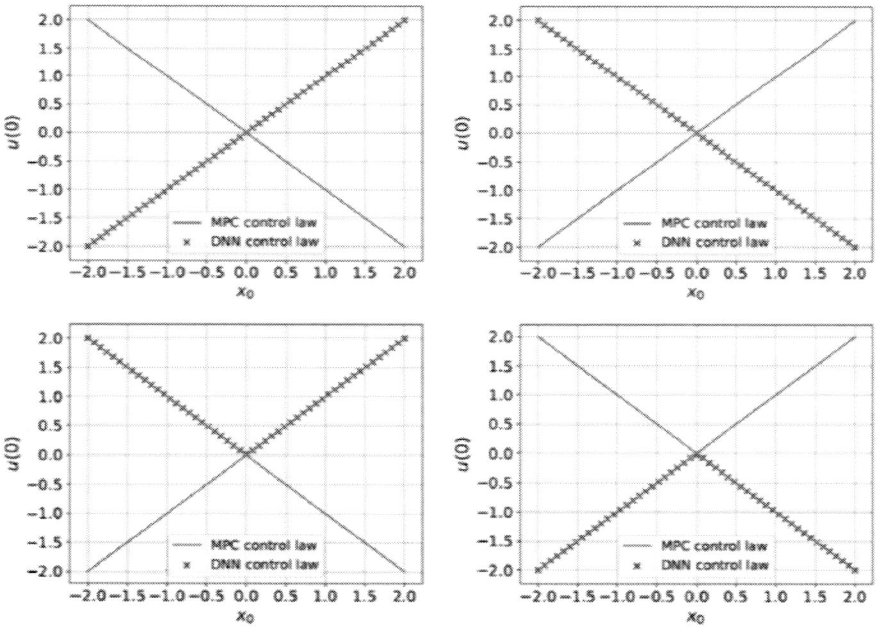

Fig. 2. DNN control law trained with "optimize and train" method

Copyright © American Institute of Chemical Engineers. All rights reserved.

601990 Cloud-Based Control of a Robotic Manufacturing Process for Personalized Medicines

Abstract:

This abstract focuses on an automated production process enabling the manufacturing of personalized drug products. Particularly, we developed a system able to fulfill the needs of small-batch, high-mix, personalized, or even on-demand, production. Recently, the concept of precision medicine has received great attention. Finding ways to manufacture personalized therapies, in a consistent and controlled way is becoming the holy grail in the pharmaceutical field. However, existing manufacturing processes are not suitable for this mass-customization approach. In fact, while traditional systems can repeatably manufacture hundreds of thousands of identical dosage forms, having being designed for one specific product at a time, personalized therapies (*i.e.* extemporaneous formulations) are produced by technicians on a small scale only. Additionally, most of these production methods require labor to monitor, regulate and complete material handling for the equipment in use.

Based on these considerations, the availability of a fully automated architecture (*i.e.* hardware and software) capable of manufacturing personalized pharmaceutical products at scale would be of utmost importance. In this respect, we have developed a modular manufacturing system and a supporting cloud-based software architecture for its control. The modular manufacturing system, referred to as "robotic cluster", consists of *i)* different manufacturing modules, each of them performing specific fabrication steps, and *ii)* a transfer system that moves work-in-progress units (WIPUs) from module to module in a distributed and parallelized way. During the manufacturing process, WIPUs are shuttled into a given module, which performs a specific step of the process. Then, the WIPUs are moved into another module, and this process is repeated for all the following operations, including quality control. The sequence of manufacturing operations for a small batch of WIPUs is performed by moving it through the modules until all needed operations are completed. This sequence can be performed in a fully-planned, fixed-scheduled, a flexible-scheduled, or dynamic manner.

Each manufacturing module is connected to a dedicated control system and to other equipment *via* its control box, enabling the input of commands for the next fabrication steps while transmitting out the data pertaining to the work that has already been performed. All critical process parameters are monitored locally and in the cloud. The strategy pursued here not only ensures the security of the manufacturing system, but also increases its versatility and flexibility. It also makes this architecture suitable for scaling, by deploying many robotic clusters at the same time. Moreover, unlike flexible manufacturing systems described in literature, ours is comprised of a set of swappable units that are readily changed, updated, repaired and maintained. This way, complex operations are divided into simpler steps and for each of them there is a corresponding custom-built module. Therefore, the principal application of this technology is in mass customization, where products can be fully personalized to each order and large quantities of products can be fabricated in parallel to achieve the same scale and throughput of traditional serial manufacturing.

As a proof of concept, we evaluated the suitability of the discussed architecture for the manufacturing of multi-compartment capsules for personalized therapy. The capsules here proposed have a modular structure that consists of multiple inner compartments, which are independent of one another. Each compartment can be filled with different active ingredients and with a range of doses, as well as formulations, for any given drug. Multi-compartment capsular devices are especially advantageous for therapies involving multiple drugs, because they simplify the dosing schedule and have a positive potential impact on overall patient compliance.

Personalized filling of each compartment of the capsule shells was performed thanks to a custom-built robotic cluster, composed of a set of automated filling modules working simultaneously, in parallel. A robotic arm (designed to safely collaborate with human operators) moves the capsular devices form one filling station to the next, and quality control machines check the deposition of the desired amount of active ingredients after each step.

Within this proof of concept, we performed an optimization on the manufacturing steps as a fixed-scheduled sequence. We created a list of variable orders *i.e.* virtual patient prescriptions of varied precise dosages of multiple drugs, and input them into a schedule optimization algorithm. Said algorithm outputs the sequence of robot arm and module actions which maximizes throughput of the cluster. Generally, this algorithm maximizes the amount of work being done in parallel at once (many batches worked on in separate modules), and schedules the order in which the cluster outputs patient-scale batches to minimize the time taken to complete all batches. To ensure quality control we bound the time between manufacturing step and quality control: in-process controls should be measured more-or-less contemporaneously with the action. No WIPU can wait for longer than 10 min - say, sitting in a buffer or idle in a module - before being processed by the corresponding quality control module. For example, we bound the time between a filling module dosing powdered formulations and a weighing module recording the dosed weight. This scheduling optimization algorithm enables the parallelism of the robotic cluster. It is through this novel planning algorithm that we achieve the same scale and throughput of traditional serial manufacturing.

The automated deposition stage allows the rapid and accurate adjustment of the combination and dosage strength of the active ingredients. Cross-contamination is avoided by enclosing each filling station in a dedicated clean room. The capsule filling systems were validated with preliminarily filling trials, which used powders with different particle size distribution and bulk density. We considered 25 batches of 50 capsule each for each powder formulation. In all cases, the amount of active ingredient deposited by the robots had an average coefficient of weight variation (*i.e.* the ratio between the standard deviation and the average value of the weight of the deposited powder) below 3%. These results demonstrate the accuracy and the flexibility of the powder deposition systems. Additionally, we use spectroscopic techniques to check the weight uniformity and identity of the pharmaceutical formulations. Every measurement is collected and saved in the cloud based quality control system, allowing the creation, in real time, of a digital master batch record with all the information relevant to each manufacturing step (*e.g.* time, date, type of machine, weight of drug deposited), which could be accessed only by verified personnel and regulatory entities.

Copyright © American Institute of Chemical Engineers. All rights reserved.

611934 Emerging Cognitive Engineering Approaches to Enhance Control Room Operator Capabilities in Chemical Industries

Abstract:

Control room operator's actions can have significant effect on the throughput, quality and safety of process plants (Baron, 1998). Typically, control room operators monitor the process via a Human Machine Interface (HMI) on which all the necessary information about the process state is presented. The complexity of operating modern chemical plants have increased significantly due to tight mass and energy integrations, deployment of sophisticated control automation strategies, and the need to operate the process in agile fashion in order to capture market and supply chain dynamics. However, there has been no concomitant increase in operations personnel's cognitive ability. Any mismatch between the operators' capability and the process control demands leads to high cognitive workload in human operators, often a precursor for poor performance which in-turn leads to sub-optimal operator actions or at worst accidents. This paper addresses the critical need to understand and enhance control room operator's cognitive capabilities.

Traditional operator performance assessment methods ignore cognitive aspects of performance. Operator's performance is assessed primarily in terms of number of successes and failures, response times, deviation of operator actions from predefined sequence (Lee et al., 2000), etc. However, in addition to overt performance (as done traditionally) it is imperative to consider covert aspects of performance i.e., cognitive workload for a holistic assessment of human performance. Various emerging cognitive engineering based sensors such as eye gaze tracking, Electroencephalography can reveal insights into the operators' cognitive state that determines these covert aspects. Specifically, in this paper we propose an Electroencephalography (EEG) based methodology to assess the operators' cognitive workload, especially during abnormal situation tasks where their active involvement is essential to safe process operations. The potential of EEG to understand cognitive workload for dynamic tasks which evolve with time has received some attention in aviation (Liang et al., 2018) and driving (Morales et al., (2017). However, there are hardly any studies relating to chemical control room operators.

We propose a single dry electrode EEG based methodology to dynamically assess the cognitive workload of operators while they perform various actions to control the process during abnormalities. The proposed methodology is illustrated on a simulated ethanol production plant testbed. A set of 10 different operators were considered in this work. Each operator's performance during the course of experiments involving monitoring and control of the process was studied. In addition to the process data and operator actions, we recorded the EEG data of the operators. Each experiment involved 6 tasks (comprising 5 different scenarios, out of which was repeated) of different abnormal situation. Each operator repeated the same experiment 7-8 times. Overall, 73 experiments were conducted by the 10 operators for a total of 438 tasks. Due to data quality issues, we evaluate results for 8 participants only.

We analyze the EEG signal in frequency domain by applying Fourier transform. Our results reveal that the power spectral density (PSD) of brain waves (4-30 Hz) is a biomarker that can identify the workload of operators. The PSD data of each operator was partitioned into two clustered using k-means clustering. Cluster 1 represents low power spectral density with low magnitude and Cluster 2 showed high power spectra density. We studied the workload variation with the help of cluster fraction over the course of the task as the operators' engagement with the process increased as well as with repetition. To evaluate the cluster variation within each task, a task is divided in three phases. In Phase 0 the operator monitors the process and ensures all the process variables are within the desired limits. During an abnormal condition an alarm is triggered, which leads to Phase 1 where the operator uses his mental model of the process to decide the control action(s) in order to clear the alarm. The portion of the task corresponding to all the control actions taken by the operator is called as Phase2. We observe distinct differences in the cluster fractions across the 3 phases. For a typical operator who successfully completes a task, during the monitoring Phase 0 we found the fraction of Cluster 1 to dominate (about 60%) indicating a correspondence between low workload during monitoring. During abnormality when alarms are triggered, in phase 1 the Cluster 1 fraction decreases (to about 40%) indicating that the operator uses his mental resources to decide the control action, i.e., increased workload. Further in phase 2, as the operator takes various control actions the Cluster 1 fraction further drops significantly (to about 0.20%) until the process becomes normal at which juncture it increases back to the 60% observed in Phase 0. The above patterns of change in the cluster pattern was observed for 81% (243 out of 300) of the tasks conducted by the 8 participants. This brings forth the capacity of the high accuracy of the EEG based cognitive sensor and the proposed machine learning algorithm to accurately estimate the workload of the control operator during normal and abnormal conditions. These distinct patterns were also highly correlated various process metrics (extent of variable deviation beyond alarm limits, time taken to clear alarm, number of standing alarms etc). In this paper, we will describe the proposed experimental and computational methodology and reported detailed results from our human subject experiments.

References:

Baron, RG. (1988). Human Factors in the Process Industries. Human Factors and Decision Making: Their Influence on Safety and Reliability, Symposium for the Safety and Reliability Society. ed. Sayers, BA. (pp.1-9)

Heard, J., Harriott, C. E., & Adams, J. A. (2018). A survey of workload assessment algorithms. IEEE Transactions on Human-Machine Systems, (99), 1-18.

Srinivasan, R., Srinivasan, B., Iqbal, M. U., Nemet, A., & Kravanja, Z. (2019). Recent developments towards enhancing process safety: Inherent safety and cognitive engineering. Computers & Chemical Engineering, 128, 364-383.

Neurosky. (2015). Greek Alphabet Soup – Making Sense of EEG Bands (Last accessed: Nov 10, 2018) http://neurosky.com/2015/05/greek-alphabet-soup-making-sense-of-eeg-bands/

Pfurtscheller, G., Stancak Jr, A., & Neuper, C. (1996). Event-related synchronization (ERS) in the α band—an electrophysiological correlate of cortical idling: a review. International journal of psychophysiology, 24(1-2), 39-46

Liang, Yongqiang, Wei Liang, Jue Qu, and Jie Yang. "Experimental Study on EEG with Different Cognitive Load." In 2018 IEEE International Conference on Systems, Man, and Cybernetics (SMC), pp. 4351-4356. IEEE, 2018.

Morales, J. M., Diaz-Piedra, C., Rieiro, H., Roca-González, J., Romero, S., Catena, A., Fuentes, L.J. & Di Stasi, L. L. (2017). Monitoring driver fatigue using a single-channel electroencephalographic device: A validation study by gaze-based, driving performance, and subjective data. Accident Analysis & Prevention, 109, 62-69.

Sharma, C., Bhavsar, P., Srinivasan, B., & Srinivasan, R. (2016). Eye gaze movement studies of control room operators: A novel approach to improve process safety. Computers & Chemical Engineering, 85, 43-57.

Bhavsar, P., Srinivasan, B., & Srinivasan, R. (2017). Quantifying situation awareness of control room operators using eye-gaze behavior. Computers & Chemical Engineering, 106,191-201.

Iqbal, M. U., Srinivasan, B., & Srinivasan, R. (2019). Electroencephalogram based Biomarkers for Tracking the Cognitive Workload of Operators in Process Industries. In Computer Aided Chemical Engineering (Vol. 46). https://doi.org/10.1016/B978-0-12-818634-3.50233-2

Copyright © American Institute of Chemical Engineers. All rights reserved.

607149 Exploring the Potential Role of Hemoglobin-Based O_2 Carriers in a Pandemic Simulation of the Blood Supply Chain

Abstract:

Introduction: The recent outbreak of coronavirus disease 2019 (COVID-19) has a significant effect on regional and national blood supply chains. Due to decreased donor availability, inadequate access to COVID-19 tests, and closure of many blood donation centers, the total number of available blood units in circulation is unable to meet clinical demand. Hemoglobin-based O_2 carriers (HBOCs) are a potential red blood cell (RBC) substitute that may be able to fill the need for RBCs during pandemics or other events that result in blood shortages. Stockpiling HBOCs for such an event is relatively easy given their long shelf life (> 6 years) at room temperature (25 °C). Although the current generation of HBOCs has similar performance compared to stored packed red blood cells (pRBCs), there may be issues related to the decreased half-life of HBOCs (1-2 days) compared to pRBCs (50 – 100 days). In light of this, the current study uses a simulation of an interregional blood supply chain in a pandemic scenario to assess what role HBOCs may have in the national blood supply.

Methods: The current study uses an interregional stock-and flow simulation of the national blood supply chain in the United States. The simulation tracks the quantity and blood type in blood banks and hospitals in 4 regions of the United States as defined by America's Blood Centers (ABC). Blood units are collected and transferred from blood banks to hospitals within their respective region. A pandemic simulation was performed by reducing the rate of blood donation within each region. Varying reduction and time for recovery were used as metrics to account for variability in the incubation period and duration of the given pathogen. HBOCs were implemented into the model as an RBC substitute that would only be used if no compatible pRBC unit is available to the patient. A half-life based algorithm was used to determine what percentage of patients would need an additional supplement of HBOC due to its relatively low half-life.

Results and Conclusion: With this model, we were able to quantify how HBOCs may help reduce strain on the blood banking system during a pandemic scenario. In these scenarios, HBOCs were able to attenuate the rate of pRBC deficits. We were also able to quantify how many units of HBOCs may be needed in future pandemic scenarios. This tool may be useful in planning and preparing for future pandemic scenarios after HBOCs obtain clinical approval.

Copyright © American Institute of Chemical Engineers. All rights reserved.

611244 Copper (II) Oxide Nanomaterial for Electroactive Inactivation of Coronavirus in Aerosols and Liquids

Abstract:

Viruses, more specifically coronaviruses, are extremely contagious and are transmitted through physical contact and airborne microdroplets. The recent SARS-CoV-2 (COVID19) has high surface stability and can remain suspended in the air for several minutes. These characteristics make disinfecting the environment difficult. Common viral disinfection methods are chemical disinfectants (chlorine and alcohol enriched products), ozonation, and ultraviolet light. These disinfecting methods can be expensive or form toxic byproducts. In this work, we present a novel approach to deactivating viruses, employing a nanostructured copper(II) oxide platform under given potentials. Prior studies have shown the inactivation of viruses on copper and copper ions surfaces[1-3]. A recent study also showed that the decay of COVID19 on copper surfaces was faster than other surfaces such as steel, plastic, and cardboard[4].

Copper(II) ions, being highly reactive, are a good candidate for inactivating viruses. The nanostructured copper(II) oxide can be integrated into various devices suitable for cleaning the environment such as a surface cleaning wand or air purifier. It can also be integrated into personal protective equipment, such as masks worn by first responders. In this work, copper foil was anodized to create copper oxide nanostructures. Various annealing conditions in oxygen were explored to create a surface enriched with copper(II), which we hypothesize will be the most effective in the inactivation of viruses[1]. Coronaviruses were introduced to the conductive copper oxide nanostructure under various potentials to induce electrochemical reactions (such as oxidizing the virus RNA). Betacoronavirus, Avian Coronavirus, and Transmissible Gastroenteritis virus were chosen as surrogates for SARS-CoV-2, tested on the cupric oxide nanostructure. The final goal of this work is to utilize the nanostructure material in an electroactive device, designed to inactive coronavirus immediately upon contact. The applied potential to the copper oxide nanostructure results in instantaneously inactivation of the virus, compared to commonly viral inactivation methods that needed extended contact time. Results on the inactivation of the mentioned viruses will be presented.

Figure (1) is an SEM picture of copper oxide nanostructures made in a mixture of , NaOH and ethylene glycol. Data on anodized copper foil in solutions containing various amounts of the mentioned mixtures such as additive of fluoride ions to get to a sufficient amount of copper(II) oxide will be presented. The oxidative state of the copper oxide nanomaterial will also be explained during the presentation by showing X-ray photoelectron spectroscopy analysis.

References:

[1] M. Horie *et al.*, “Inactivation and morphological changes of avian influenza virus by copper ions,” *Arch. Virol.*, vol. 153, no. 8, pp. 1467–1472, 2008.

[2] Sundberg, K.; Champagne, V.; McNally, B.; Helfritch, D.; Sisson, R., Effectiveness of nanomaterial copper cold spray surfaces on inactivation of influenza A virus. J Biotechnol Biomater 2015, 5 (205), 2.

[3] Longano, D.; Ditaranto, N.; Sabbatini, L.; Torsi, L.; Cioffi, N., Synthesis and antimicrobial activity of copper nanomaterials. In Nano-Antimicrobials, Springer: 2012; pp 85-117.

[4] D. Taylor, A. C. Lindsay, and J. P. Halcox, "c or r e sp ondence Aerosol and Surface Stability of SARS-CoV-2 as Compared with SARS-CoV-1," pp. 0–2, 2010.

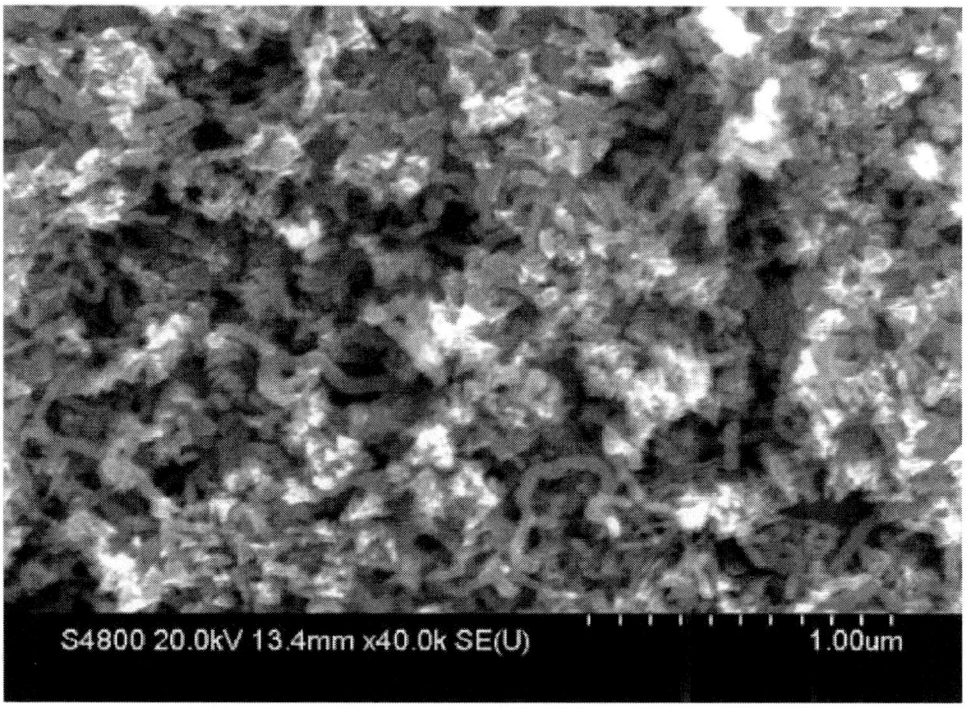

Figure (1), A SEM image of synthesized copper oxide sensor.

Copyright © American Institute of Chemical Engineers. All rights reserved.

598367 Application of Biophysical and Chemical Engineering Principles for Understanding Molecular Scale Interactions Critical to Virus Entry and Infection of Its Host

Abstract:

The coronavirus disease 2019 (COVID-19) pandemic has focused attention on the need to develop effective therapies against the causative agent, SARS-CoV-2, and also against other pathogenic coronaviruses (CoV) that have emerged in the past or might appear in future. Focusing on steps in the CoV replication cycle, in particular the entry steps involving membrane fusion that are vulnerable to inhibition by broad-spectrum or specific antiviral agents, is an astute choice because of the conserved nature of the fusion machinery and mechanism across the CoV family. For coronavirus, entry into a host cell is mediated by a single glycoprotein protruding from its membrane envelope, called spike (S). Within S, the region that directly interacts with the membrane is called the fusion peptide, FP. It is the physico-chemical interactions of the FP with the host membrane that anchors it, thus enabling the necessary deformations of the membrane that lead to delivery of the viral genome into the cell when a fusion pore opens. As chemical engineers, we contribute to this fundamental work by leveraging our understanding of thermodynamics, kinetics, and intermolecular interactions to describe FP interactions with the host membrane at the most fundamental molecular level to facilitate the development of strategies to limit those interactions to stop the spread of infection. In this talk, I will describe our work on understanding the impact of calcium ions on CoV infection. Using cell infectivity, biophysical assays, and spectroscopic methods, we found that calcium ions serve to stabilize the fusion peptide structure during conformational change that then allows its insertion into the host membrane, resulting in increased lipid ordering in the membrane. This lipid ordering precedes membrane fusion and has been shown to correlate with increased fusion activity, as higher extents of fusion are observed as calcium concentration increases, aligned with higher levels of infection in the presence of calcium. Finally, depletion of calcium ions leads to structure and activity changes that correlate well with *in vitro* experiments using calcium-chelating drugs. Under these conditions, cell infection dropped, pointing to the possibility of such drugs as therapeutic interventions.

Copyright © American Institute of Chemical Engineers. All rights reserved.

605066 Multiscale Lung Tissue Simulator for Sars-Cov-2 Infection and Damage

Abstract:

The 2019 novel coronavirus, SARS-CoV-2, is an emerging pathogen of critical significance to international public health. Knowledge of the interplay between molecular-scale virus-receptor interactions, single-cell viral replication, intracellular-scale viral transport, and emergent tissue-scale viral propagation is limited. Moreover, little is known about immune system-virus-tissue interactions and how these can result in low-level (asymptomatic) infections in some cases and acute respiratory distress syndrome (ARDS) and other tissue damage in others, particularly with respect to presentation in different age groups or pre-existing inflammatory risk factors like diabetes. A critical question for treatment and protection is why it appears that the severity of infection may correlate with the initial level of virus exposure. Given the nonlinear interactions within and among each of these processes, multiscale simulation models can shed light on the emergent dynamics that lead to divergent outcomes, identify actionable "choke points" for pharmacologic interactions, screen potential therapies, and identify potential biomarkers that differentiate patient outcomes. Given the complexity of the problem and the acute need for an actionable model to guide therapy discovery and optimization, we have introduced a prototype of a multiscale mathematical model of SARS-CoV-2 dynamics in lung and intestinal tissue that will be iteratively refined. The first prototype model was built and shared as open source, cloud-hosted and web-executed code and simulator[1] in under 12 hours in March 2020. We have assembled an international, multi-disciplinary collaborative team called the SARS-CoV-2 Tissue Simulation Coalition with academic, industry, and non-profit sector participants. Additional domain experts have continued to join the Coalition throughout its first month. We have documented the rapid model development process with a preprint submission[2], and second version is to be released within a few days. In a sustained community effort, this model is integrating data and expertise across virology, immunology, mathematical biology, quantitative systems physiology and pharmacology, cloud and high performance computing, and other domains to accelerate our response to this critical health threat.

We are developing an open-source, multi-scale tissue simulator that can be used to investigate mechanisms of intracellular viral replication, infection of epithelial cells, host immune response, and tissue damage. The aim of this project is to concentrate community modeling efforts to create a comprehensive multiscale simulation framework that can subsequently be calibrated, validated, and used to rapidly explore and optimize therapeutic interventions for COVID-19. Once the prototype has been completed (after several design iterations), this coalition will transition to maintain and support the simulation and data collection/curation framework and aggregate calibrated/validated parameter values. To address the acute need for rapid access to an actionable model, we are using a community-driven coalition and best open science practices to build and iteratively refine the model.

Multiple chemical engineers are in the SARS-CoV-2 Tissue Simulation Coalition. Even from the first prototype version, the model applies chemical engineering fundamentals to systemically analyze dynamic interactions between processes involved in SARS-CoV-2 infection and damage. Chemical reactions, enzyme kinetics, and molecular binding interactions are simulated for viral and host cellular processes. Transport phenomena including extracellular diffusion of secreted chemicals and cell endocytosis and exocytosis of viral materials are explicitly included in the current version. In future versions, we plan to consider convection of aerosol droplets laden with virus throughout the lungs, penetration of the virus through the mucus lining the airways, signaling to systemic host responses, and distribution of systemic immune responses into local tissues. Core concepts in mass and population balances, numerical methods, and mathematical modeling with ordinary and partial differential equations are also used throughout the project. Stochastic and multiscale modeling approaches are also incorporated to address challenges at different length and times scales and to allow for uncertainty and biological variations. This is not simply a theoretical and computational study. Experimental and clinical collaborators who are setting up for animal and cell culture experiments in April 2020 are also part of the Coalition. It is anticipated that these data as they become available will be incorporated into multiple versions of the model before the AIChE Annual Meeting.

1. Heiland, R.; Macklin, P. PhysiCell model for COVID19. https://nanohub.org/tools/pc4covid19
2. Wang, Y. N.; Heiland, R.; Craig, M.; Davis, C.; Ford Versypt, A. N.; Jenner, A.; Ozik, J.; Collier, N.; Cockrell, C.; Becker, A.; An, G.; Glazier, J. A.; Narayanan, A.; Smith, A. M.; Macklin, P. *bioRxiv* **2020,** 2020.04.02.019075, https://doi.org/10.1101/2020.04.02.019075.

Copyright © American Institute of Chemical Engineers. All rights reserved.

605590 Inflammation Product Effects on Dilatational Mechanics Can Trigger the Laplace Instability and Acute Respiratory Distress Syndrome

Abstract:

In the lungs, the Laplace pressure, $\Delta P = 2\gamma/R$, would be higher in smaller alveoli than larger alveoli unless the surface tension, γ decreases with alveolar interfacial area, A, such that $2\varepsilon > g$ in which $\varepsilon = A(d\gamma/dA)$ is the dilatational modulus. In Acute Respiratory Distress Syndrome (ARDS), lipase activity due to the immune response to an underlying trauma or disease causes single chain lysolipid concentrations to increase via hydrolysis of double-chain phospholpids. Increasing lysolipid concentrations decrease the dilatational modulus dramatically at breathing frequencies if the soluble lysolipid has sufficient time to diffuse off the interface, causing $2\varepsilon < \gamma$, thereby inducing the "Laplace Instability", in which larger alveoli have a lower internal pressure than smaller alveoli. This can lead to uneven lung inflation, alveolar flooding, and poor gas exchange, typical symptoms of ARDS, in particular ARDS associated with Covid-19 infections. At high frequencies, even at high lysolipid concentrations, $2\varepsilon - \gamma > 0$, which explains the benefits ARDS patients receive from high frequency oscillatory ventilation.

Copyright © American Institute of Chemical Engineers. All rights reserved.

611834 Elucidating Molecular Origins of Drug Selectivity for Rational Design of Selective Drugs

Abstract:

Proteins families that evolve to bind particular ligand also contain diverse members that exhibit a wide range of affinities for that particular ligand. However, this variance in affinity resulting from evolution of the family members is usually at odds with the high sequence and structure similarity shared by its members. In this talk, we discuss three such protein families i.e. human protein kinases, Cannabinoid receptors and Strigolactone receptors in plants. These proteins show several orders of magnitude difference in their ligand affinity but share high sequence and structural similarity. The characterization of the free energy landscapes for ligand binding starting could shed light on the molecular origin of their ligand selectivity mechanisms. In order to achieve this goal, we have develop computational methods for evaluation of their binding mechanism using molecular simulations. These examples served as model system for obtaining generalizable insights about the free energy landscapes of ligand binding that could help design of drugs with desired selectivity profiles.

Copyright © American Institute of Chemical Engineers. All rights reserved.

600211 Computer Aided Generation of Diverse and Synthesizable Molecular Library

Abstract:

Modern drug discovery is extending into chemical spaces with increased complexity. It is desirable to explore these spaces more efficiently. In order to maximize the information gained from testing new molecules, the concept of molecular library has been widely adopted in medicinal chemistry.

Molecular libraries are collections of compounds that have, or expected to have, similar structures or properties. Accessing all molecules in a molecule library can enable the rapid generation of structure-activity relationships during the hit-to-lead or lead optimization phases, which can greatly enhance the efficiency of drug discovery. Small molecule libraries can be constructed in many different ways. A library may be a collection of compounds extracted from the literature with similar functionality;[1,2] or it may be designed by enumerating possible side groups as decorations of a common core scaffold;[3–5] it may also be designed using any one of the increasing number of *in silico* tools for the generation of drug-like compound libraries.[6–8]

Despite the significant progress in molecular library generation, there is a usually a tradeoff between accessibility and diversity – libraries that are constructed by simply substituting functional groups on the same scaffold are usually limited in diversity, and libraries with larger structural diversity tend not to be all accessible through similar reactions. In this work, we leverage informatics tools to maximize the diversity of molecular libraries as well as their synthetic accessibility. Starting from a given molecule, using retrosynthesis analysis and reaction prediction models, we can generate libraries that can be accessed through alternative late-stage functionalization. The diversity of the molecular libraries is assessed by average pairwise similarity, so that the most diverse molecular library can be identified. This work provides insight on expanding the exploration of surrounding chemical space of a molecule with minimal additional effort.

1 U. F. Röhrig, S. R. Majjigapu, P. Vogel, V. Zoete and O. Michielin, *J. Med. Chem.*, 2015, **58**, 9421–9437.

2 B. T. Xin, E. M. Huber, G. De Bruin, W. Heinemeyer, E. Maurits, C. Espinal, Y. Du, M. Janssens, E. S. Weyburne, A. F. Kisselev, B. I. Florea, C. Driessen, G. A. Van Der Marel, M. Groll and H. S. Overkleeft, *J. Med. Chem.*, 2019, **62**, 1626–1642.

3 K. Kannan Sivaraman, A. Paiardini, M. Sieńczyk, C. Ruggeri, C. A. Oellig, J. P. Dalton, P. J. Scammells, M. Drag and S. McGowan, *J. Med. Chem.*, 2013, **56**, 5213–5217.

4 R. Fleeman, T. M. Lavoi, R. G. Santos, A. Morales, A. Nefzi, G. S. Welmaker, J. L. Medina-Franco, M. A. Giulianotti, R. A. Houghten and L. N. Shaw, *J. Med. Chem.*, 2015, **58**, 3340–3355.

5 G. M. Keseru, D. A. Erlanson, G. G. Ferenczy, M. M. Hann, C. W. Murray and S. D. Pickett, *J. Med. Chem.*, 2016, **59**, 8189–8206.

6 R. Gómez-Bombarelli, J. N. Wei, D. Duvenaud, J. M. Hernández-Lobato, B. Sánchez-Lengeling, D. Sheberla, J. Aguilera-Iparraguirre, T. D. Hirzel, R. P. Adams and A. Aspuru-Guzik, *ACS Cent. Sci.*, 2018, **4**, 268–276.

7 M. H. S. Segler, T. Kogej, C. Tyrchan and M. P. Waller, *ACS Cent. Sci.*, 2018, **4**, 120–131.

8 A. Gupta, A. T. Müller, B. J. H. Huisman, J. A. Fuchs, P. Schneider and G. Schneider, *Mol. Inform.*, 2018, **37**, 1700111.

Copyright © American Institute of Chemical Engineers. All rights reserved.

605512 *De Novo* Design of Engineered Cytokine Mimetics As Targeted Immunotherapeutics.

Abstract:

Cytokines are soluble factors that signal through stimulation of their cognate transmembrane receptors on target cells to perform critical biological functions, particularly those related to immune homeostasis. As a result of their essential immune activities, cytokines have great potential as immunotherapeutics, both to activate the immune response to fight diseases such as cancer and infectious diseases, as well as to suppress the immune response to treat autoimmune disorders or for transplantation medicine. However, the pleiotropic activities and unfavorable pharmaceutical properties of natural cytokines have limited their clinical performance.

Recent efforts to re-engineer native cytokines to improve their selectivity and stability have significantly advanced progress in cytokine therapeutic development. However, there are inherent challenges to developing therapeutics from natural proteins. First, most natural cytokines are only marginally stable, hence mutations aimed at increasing efficacy can decrease expression or cause aggregation, making manufacturing and storage difficult. More substantial changes, such as the deletion or fusion of functional or targeting domains, can also dramatically alter pharmacokinetic properties and tissue penetration. In addition, immune responses targeting a variant of a natural protein may cross-react with the endogenous molecule with potentially catastrophic consequences. Finally, the high potency and multifarious functions of cytokines hinder their efficacy and can result in harmful off-target effects and systemic toxicities.

Here, we combine a best-in-class computational protein design software with directed evolution technologies to generate *de novo* cytokine mimetics that circumvent limitations for naturally-derived cytokine drugs. Rather than modifying existing molecules, we adopt a bold new approach to create exceptionally stable proteins with customized receptor interaction properties as robust and efficacious targeted therapeutics. We applied this strategy to develop a mimetic of the interleukin-2 (IL-2) cytokine, which plays a pivotal role in T and natural killer (NK) cell function. We created a hyper-stable *de novo* protein that is biased toward activation of IL-2 signaling pathways on immune effector cells, and showed that this molecule inhibits tumor growth and prolongs survival in mouse models of colorectal cancer and melanoma, while also mitigating the toxic side effects typically associated with IL-2 therapy. We built upon this exciting advance to design *de novo* cytokines that mimic IL-4, which promotes pro-regenerative immune programs following tissue damage. We demonstrated the extreme thermal stability and biased receptor activity of our engineered IL-4 mimetic, and showed that it elicited pro-regenerative immune responses in both cellular and animal models. Collectively, our work pioneers a novel cytokine engineering platform that integrates computational and experimental approaches to establish a general paradigm for advancing the clinical translation of cytokine therapeutics.

Copyright © American Institute of Chemical Engineers. All rights reserved.

606100 Integrated Prediction of Chemical Effects on Important Pathways of Embryo Development Based on in Vitro Embryonic Stem Cell Test and Machine Learning

Abstract:

Currently, there are thousands of chemicals in common use; however, only a small number of them have sufficient toxicity evaluation, and even less information for the specific embryo-related toxicity field: embryotoxicity and developmental toxicity. Embryonic stem cell test (EST) is the only accepted *in vitro*method for assessing embryotoxicity without animal sacrifice. However, EST for regulatory embryotoxicity screening are impeded by its complexity, time-consuming, limited differentiation lineage endpoints, and poor availability for use to develop a reliable prediction model (PM) for embryotoxicity assessments.

A key step in the understanding of embryo development is to construct a systematic and comprehensive model of network with the development-related pathways involved. In our research, we mainly focus on the essential markers in the pathways related to ESCs apoptosis and proliferation, pluripotency, and specific-lineage differentiation. Quantitative structure-activity relationship (QSAR) models which predict the biological activities from chemical structures information is an efficient and less time-consuming strategy compared to experimental-based studies. Therefore, an integrated prediction pipeline for *in vitro*high throughput screening (HTS) based on engineered murine embryonic stem cells (mESCs) expressing a reporter, enhanced green fluorescent protein (EGFP), driven by human promoters for key genes in the pathways associated with apoptosis, pluripotency, and tissue-specific differentiation cultured in novel multi-well plates is developed and used to obtain new embryotoxicity and developmental toxicity data needed for the training of machine learning models such as support vector machines (SVM) and deep neural network (DNN). The application of advanced machine learning algorithms can significantly improve the accuracy and efficiency of the PM for embryotoxicity assessment of compounds with unknown developmental toxicity. This study will provide the technology needed for fast and accurate embryotoxicity assessments. The data obtained from our integrated analysis can provide the information needed for the regulation of industrial and emerging chemicals.

Copyright © American Institute of Chemical Engineers. All rights reserved.

607548 A Computational Approach to Identifying Synergistic Compounds for Treating Antimicrobial-Resistant Pathogens

Abstract:

Antimicrobial resistance (AMR) is increasingly a threat to global public health [1]. AMR-related infections limit the efficacy of lifesaving antibiotic therapy necessary for the treatment of infectious disease and limit the success of advanced surgical procedures like organ transplants [2, 3]. To combat AMR pathogens, either new antibiotics or the cocktails of existing antibiotics with inhibitors of antimicrobial resistance proteins should be explored. Since the development of new antibiotic has been slowed down significantly recently, identifying compounds/antimicrobials with synergistic effects becomes the major approach to addressing the AMR crisis. In this work, we demonstrate a computational approach to identifying plant compounds that have synergistic effect with Fosfomycin to inhibit *Listeria monocytogenes*, a food-borne pathogen that could cause listeriosis disease, especially on immune-compromised people [4, 5]. Experimental results further validate the our finding.

L. *monocytogenes* was found to infect and adversely affect patients' liver and spleen. In addition, *L. monocytogenes* can penetrate blood-brain barrier and blood-placenta barriers to harm the central neural system of pregnant woman and infant [4]. Since *L. monocytogenes* is an intracellular pathogen, antimicrobials used to treat listeriosis should be able to be transported into host cells. Penicillin, ampicillin and amoxicillin were commonly used antibiotics in the treatment of listeriosis [6]. However, antibiotic resistance genes have been continuously found in *Listeria* strains. For example, the strain that has resistance to Penicillin G was isolated from vegetables in 2016 [4, 7]. As a natural product, Fosfomycin was found effective against clinical isolates of *L. monocytogenes* and used as a novel therapeutic antibiotic for listeriosis clinical treatment [8]. In addition, Fosfomycin is able to penetrate the blood-brain barrier and reach clinically relevant concentrations. Thus, it has the potential to eliminate *L. monocytogenes* which would cause neuron damage[9]. However, stronger Fosfomycin resistance was found in the *L. monocytogenes* isolates with Fosfomycin resistance proteins detected [10]. In particular, a resistant gene FosX (LMO1702,402bp) was identified and expressed in *L. monocytogenes* EGDe (strain ATCC BAA-679), a typical well-studied strain [11]. The FosX enzyme catalyzed the hydrolysis of Fosfomycin and resulted in the Fosfomycin resistance in *L. monocytogenes* EGDe [5, 10]. Therefore, there is an urgent need to identify compounds that can inhibit FosX enzyme to revive the efficacy of Fosfomycin to treat *L. monocytogenes.*

Automated molecular docking is the most commonly used computational approach that evaluates the binding of small-molecule ligands like compounds to a target receptor with a known crystal 3D structure [12]. Molecular docking provides an avenue for a high-throughput virtual screening of ligands, and it has been widely implemented in drug discovery research for hit identification[13]. Docking programs have been improved recently to provide more accurate prediction on ligand-target binding by optimizing docking algorithms and scoring functions [14]. Among those existing docking programs, Molsoft ICM was evaluated with 93% accuracy in flexible docking and 90% successful rate in covalent docking. This was significantly better than the performance of Autodock, DOCK, FlexX, Gold, FITTED and MOE [13-15]. Since structures and activities of the FosX protein in *L. monocytogenes* have been well studied, we used an integrated ICM-docking and experimental approach to identify FosX inhibitors that are of synergistic effect with Fosfomycin in treating resistant *L. monocytogenes*. Specifically, automated ligand docking was implemented to perform virtual screening of Indofine natural-product database and FDA-approved drugs to identified potential inhibitors. *In vitro* bacterial growth inhibition test was then utilized to verify the effectiveness of identified compounds combined with Fosfomycin on inhibiting the resistant *L. monocytogenes*. We demonstrated that two phenolic acids, i.e., caffeic acid and chlorogenic acid, were predicted as high-affinity FosX inhibitors from the ligand-docking platform. Experiments with these compounds indicated that: the cocktail of either caffeic acid (1.5mg/mL) or chlorogenic acid (3mg/mL) with Fosfomycin (50mg/L) was able to significantly inhibit the growth of the pathogen [16]. The finding of this work implies that the combination of Fosfomycin with either caffeic acid or chlorogenic acid is of potential to be used in the clinical treatment of *Listeria* infections.

References

1. Velez, R. and E. Sloand, *Combating antibiotic resistance, mitigating future threats and ongoing initiatives.* J Clin Nurs, 2016. **25**(13-14): p. 1886-9.
2. Michael, G.B., et al., *Emerging issues in antimicrobial resistance of bacteria from food-producing animals.* Future Microbiol, 2015. **10**(3): p. 427-43.
3. Davies, J. and D. Davies, *Origins and evolution of antibiotic resistance.* Microbiol Mol Biol Rev, 2010. **74**(3): p. 417-33.
4. Radoshevich, L. and P. Cossart, *Listeria monocytogenes: towards a complete picture of its physiology and pathogenesis.* Nat Rev Microbiol, 2018. **16**(1): p. 32-46.
5. Scortti, M., et al., *Epistatic control of intrinsic resistance by virulence genes in Listeria.* PLoS Genet, 2018. **14**(9): p. e1007525.
6. Temple, M.E. and M.C. Nahata, *Treatment of Listeriosis.* Annals of Pharmacotherapy, 2000. **34**(5): p. 656-661.
7. de Vasconcelos Byrne, V., et al., *Occurrence and antimicrobial resistance patterns of Listeria monocytogenes isolated from vegetables.* Brazilian Journal of Microbiology, 2016. **47**(2): p. 438-443.
8. Lepe, J.A., et al., *In vitro and intracellular activities of fosfomycin against clinical strains of Listeria monocytogenes.* International Journal of Antimicrobial Agents, 2014. **43**(2): p. 135-139.
9. Falagas, M.E., et al., *Fosfomycin.* Clin Microbiol Rev, 2016. **29**(2): p. 321-47.
10. Fillgrove, K.L., et al., *Mechanistic diversity of fosfomycin resistance in pathogenic microorganisms.* J Am Chem Soc, 2003. **125**(51): p. 15730-1.
11. Fillgrove, K.L., et al., *Structure and mechanism of the genomically encoded fosfomycin resistance protein, FosX, from Listeria monocytogenes.* Biochemistry, 2007. **46**(27): p. 8110-20.
12. Meng, X.Y., et al., *Molecular docking: a powerful approach for structure-based drug discovery.* Curr Comput Aided Drug Des, 2011. **7**(2): p. 146-57.
13. Kitchen, D.B., et al., *Docking and scoring in virtual screening for drug discovery: methods and applications.* Nat Rev Drug Discov, 2004. **3**(11): p. 935-49.
14. Lionta, E., et al., *Structure-Based Virtual Screening for Drug Discovery: Principles, Applications and Recent Advances.* Current Topics in Medicinal Chemistry, 2014. **14**(16): p. 1923-1938.
15. Scarpino, A., G.G. Ferenczy, and G.M. Keserű, *Comparative Evaluation of Covalent Docking Tools.* Journal of Chemical Information and Modeling, 2018. **58**(7): p. 1441-1458.
16. Zhang, F., et al., *Synergistic Effect of Chlorogenic Acid and Caffeic Acid with Fosfomycin in Growth Inhibition of a Resistant Listeria monocytogenes Strain.* ACS Omega, 2020. **https://dx.doi.org/10.1021/acsomega.0c00352**.

Copyright © American Institute of Chemical Engineers. All rights reserved.

594532 Liquid Phase Membrane Supported Cyclical Flow Synthesis of Oligonucleotides for Therapeutic Applications

Abstract:

There now exists a consensus that RNA based gene inference techniques such as Anti-Sense Oligonucleotides and siRNA represent a significant therapeutic modality in the pharmaceutical industry. While the applications and understanding of the underlying biology of these systems for use as therapeutic platforms have advanced rapidly, little change in the commercial synthesis of these compounds has occurred. Metric tonnes per annum of oligonucleotides are projected to be required due to the high number of oligonucleotide-containing drugs in late phases of clinical trials and also projected for individual ASO therapies as larger patient populations are targeted in clinical trials following a number of clinical successes in small orphan patient populations.[1][2] The existing state-of-the-art synthesis of oligonucleotides is carried out with the use of a solid phase support and batch sizes currently limited to approximately 5 kg due to the constraints of solid phase synthesis.[3] In contrast, liquid phase supports reduce many of the limitations that are associated with solid phase supports.[4] Liquid phase synthesis of oligonucleotides was established by using linear and branched polyethylene glycol (PEG) as the soluble support[5] Membrane enhanced synthesis of oligonucleotides and peptides using PEG as a soluble support have been shown to be a potential solution for the problems associated with solid phase synthesis.[6][7]

The current work demonstrates proof of concept for a novel liquid phase manufacturing process utilizing cyclical flow synthesis combined with a novel membrane separation (**Figure 1**). This should enable fully scalable cyclical flow synthesis of oligonucleotides with increased productivity within a compact and scalable reactor footprint. This novel process intends to meet the growing demands of the industry and promises reduced capital costs and increased product quality control.

Modifications to soluble supports were utilized to both optimize the nucleotide loading process but also to optimize membrane flux and rejection factor, and hence yield and purity with commercially available membranes. Both liquid batch and phase cyclical flow synthesis of short oligonucleotides were optimized for coupling yield, phosphoramidite consumption, solvent consumption and overall productivity. The ability to use optimum reaction/separation timescales which are inherent to coupling reactions in oligonucleotide synthesis enabled improved system performance and in turn the ability to rapidly supply oligonucleotides on demand.

Acknowledgements:

This publication is supported by Science Foundation Ireland (SFI) through SSPC, The SFI Research Centre for Pharmaceuticals (Grant number 12/RC/2275_P2).

[1] Y. S. Sanghvi, *Curr. Protoc. Nucleic Acid Chem.* **2011**, 1–22.

[2] A. G. Molina, Y. S. Sanghvi, *Curr. Protoc. Nucleic Acid Chem.* **2019**, *77*, 1–17.

[3] A. R. Lajmi, L. Schwartz, Y. S. Sanghvi, *Org. Process Res. Dev.* **2004**, *8*, 651–657.

[4] P. R. J. Gaffney, J. F. Kim, I. B. Valtcheva, G. D. Williams, M. S. Anson, A. M. Buswell, A. G. Livingston, *Chem. - A Eur. J.* **2015**, *21*, 9535–9543.

[5] B. Burcovich, F. M. Veronese, V. Zarytova, G. M. Bonora, in *Nucleosides and Nucleotides*, Marcel Dekker Inc., **1998**, pp. 1567–1570.

[6] M. Mutter, H. Hagenmaier, E. Bayer, *Angew. Chemie* **1971**, *83*, 883–884.

[7] F. Brandstetter, H. Schott, E. Bayer, *Tetrahedron Lett.* **1973**, *14*, 2997–3000.

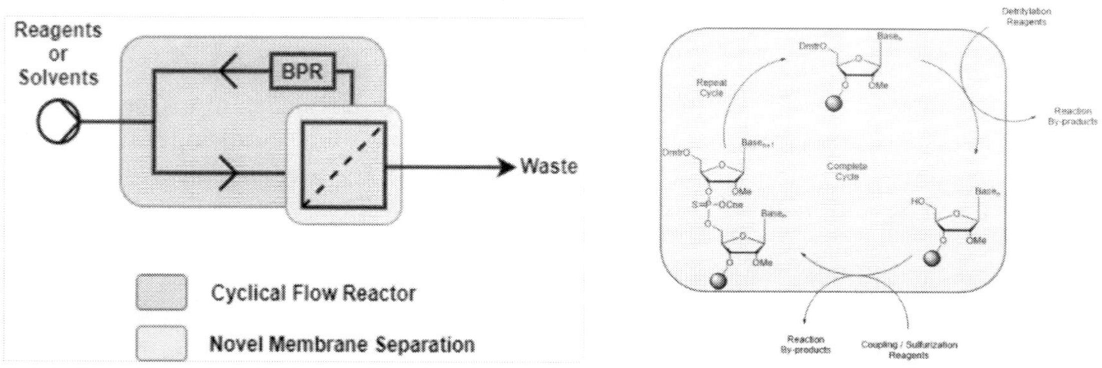

Figure 1: Schematic of system configuration (left) and Overview of cyclical synthetic route (right)

Copyright © American Institute of Chemical Engineers. All rights reserved.

605370 Predict New Cocrystals Via Mechanochemistry

Abstract:

Within the pharmaceutical industry around 60% of recently developed drugs have solubility challenges while 80% of in-production line drugs are classified as being poorly soluble. Solubility is a critical property for drug performance (e.g. bioavailability) [1,2]. Cocrystal formation is one possible solution to overcome solubility problems. Pharmaceutical cocrystals, formed by combining an active pharmaceutical ingredient (API) with a cocrystal former (coformer), are often manufactured in a variety of methods that rely on the utilization of solvents. Pharmaceutical companies use between 25-100 kg solvent per kg of product, causing major water and air pollution. Hence, state-of-the-art solvent-based methods need to be replaced with more sustainable alternatives. Additionally, the formation of a cocrystal requires its components to be soluble in the same solvent. These limitations hinder the processability and development of novel pharmaceutical products.

Mechanochemistry utilizes mechanical forces to induce reactions. The application of mechanochemistry to form cocrystals without solvents is a novel, promising and sustainable method [1,3]. Scientists have exploited the features of mechanochemistry, but only in a random, non-systematic manner for the production of new cocrystals. Indeed, with the enormous number of potential API-coformer combinations, it is currently impossible to predict if two crystalline materials will cocrystallize under mechanochemical forces.

In this work, a high throughput screening approach combined with data analysis (machine learning) has been developed to predict if two crystals will cocrystallize under mechanochemical conditions. Strategically distinct pairs of APIs and crystalline coformers were screened by neat grinding (solventless) in a Retsch MM 400 vibration-mill. Powder X-ray diffraction has been applied to characterize all the samples and to determine the formation of new crystal structures. With the help of these results and molecular descriptors a machine learning algorithm (random forest model) has been successfully developed, effectively predicting the propensity of two molecules to cocrystallize under mechanochemical conditions. Additional cocrystallization events are predicted and experimentally validated. The technical feasibility of further API-coformer combinations can be estimated to create novel cocrystals without any solvent limitations, expanding the scope of mechanochemistry as a technique for drug discovery.

[1] C. B. Aakeröy and A. S. Sinha. Co-crystals. Monographs in Supramolecular Chemistry. The Royal Society of Chemistry, 2018.
[2] M. A. E. Yousef and V. R. Vangala. Pharmaceutical cocrystals: Molecules, crystals,formulations, medicines. Crystal Growth & Design, 19(12):7420–7438, 2019.
[3] S. Lou, Y. Mao, D. Xu, J. He, Q. Chen, and Z. Xu . Fast and Selective Dehydrogenative C–H/C–H Arylation Using Mechanochemistry. ACS Catalysis, 6 (6) , 3890-3894, 2016.

Copyright © American Institute of Chemical Engineers. All rights reserved.

606293 The SIR Dynamic Model of Infectious Disease Transmission and Its Analogy with Chemical Kinetics

Abstract:

The classic Susceptible-Infectious-Recovered (SIR) mathematical model of the dynamics of infectious disease transmission resembles a dynamic model of a batch reactor carrying out an autocatalytic reaction with catalyst deactivation. This analogy between disease transmission and chemical reactions allows chemists and chemical engineers to peer into dynamic models of infectious disease transmission used to forecast epidemics and assess mitigation strategies. Moreover, analysis of SIR model dynamics gives insights into the kinetics of autocatalytic reactions.

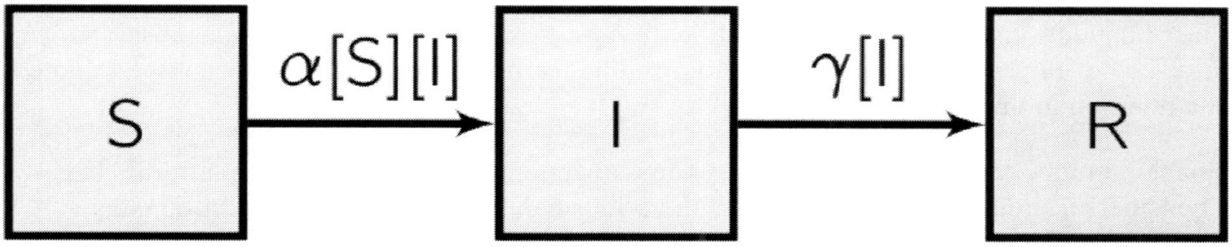

Copyright © American Institute of Chemical Engineers. All rights reserved.

606629 A Conceptual Study on the Epidemic Spread Based on Cellular Automata and Feature Extraction

Abstract:

With the progress in process systematic theory and modelling practice, more and more social and biological system can be studied as a generalized process system, the spread of infectious disease can fall into the category of such systems, which involve emergence of initial infection, diffusion-like propagation across the geographic area and time, and have been mathematically studied since the beginning of last century. Amongst various mathematical methods for describing the spread of infectious diseases, cellular automata make it feasible to explicitly simulate the spatial and temporal evaluation of the epidemic with intuitive local rules obtained from the mechanism of spread for a specific disease. Unfortunately, these local rules are usually covered by disturbed symptoms and suffering of patients, and are available to public through a long-term epidemiological observation and analysis. The development of data mining and machine learning techniques significantly facilitate the feature extraction in this area.

In this paper, a multi-scale model is proposed and realized on a cellular automata platform. This model considers a simplified social system with various parameters, including infection rates, infection and quarantine period, population movement, as well as the citizen awareness. When all these factors exist at the same time and interact over time in the system, the critical factors for epidemic control become diverse and unpredictable, which makes the situation even more challenge for healthcare professionals and decision makers. With proper data-driven feature extraction method, the critical factors for epidemic spread can be identified for further situation control.

All the impact factors mentioned above are tested by adjusting the corresponding parameters in this model, so that the quantitative influence of those factors are investigated respectively. Based on our results, an average incubation period hidden in the model mechanism can be identified by data analysis, which is key information for the development of quarantine strategies, and the intergenerational evolution of virus can be obtained through infectivity decay to understand the current status of epidemics and obtain a reference for adjusting the clinic treatment and control strategy accordingly.

Copyright © American Institute of Chemical Engineers. All rights reserved.

611126 Stochastic Progression of COVID-19 Symptoms

Abstract:

COVID-19 is a pandemic viral disease endangering a yet unknowable number of individuals. This disease is two to three times more contagious than influenza, and it is transmissible before symptoms appear. As a result, contact with infected individuals leads to cluster outbreaks. If patients with suspected cases were quickly tested, isolated and instructed to notify their recent contacts, these outbreaks could be more effectively contained. Unfortunately, patients with COVID-19 might have symptoms similar to other common illnesses. Thus, understanding the onset of COVID-19 symptoms might help instruct the public when to seek medical attention. To this end, we apply a Markov Process to a graded partially ordered set based on clinical observations of confirmed cases of COVID-19 to ascertain the most likely order of discernible symptoms (i.e. fever, cough, nausea/vomiting, and diarrhea) in COVID-19 patients. We then compared the progression of these symptoms in COVID-19 to other respiratory diseases, such as influenza, SARS and MERS, to observe if the disease presents differently. Our calculated predictions are consistent with the notion that influenza initiates with cough, whereas COVID-19 is similar to other coronavirus-related diseases, SARS and MERS, in initiating with fever. However, COVID-19 differs from SARS and MERS in the order of gastrointestinal symptoms. Our results support the recommended early checking method for COVID-19 from the CDC, which advises the public to take their temperature at home and when entering facilities. Additionally, our findings suggest that clinicians should consider recording the order of symptoms in COVID-19 patients, and if confirmed with prospective clinical data, this information could reduce the outbreak.

Copyright © American Institute of Chemical Engineers. All rights reserved.

Modeling Clinical Software for Implementing A RAPID Response Public Health Safety and Treatment DATA Management System for NOVEL Coronavirus-Convid 19 Pandemic

Dr Kingsley Eromoses Abhulimen
Department of Chemical Engineering
University of Lagos, Nigeria

University Technology System Limited
Syntechsys Corporation TX77077
United States of America
syntechsysglobal@gmail.com

ABSTRACT

This paper presents the clinical model algorithms and MATLAB software simulations of CONVID-19 viral antibodies mapping in human DNA immune system for implementing diagnosis and analysis of clinical data obtained from Lagos hospital in Lagos State, Nigeria. The model used simulated viral epidemic setting with published literature data of Wuhan China CONVID-19 patients admitted into the Intensive care unit (ICU) to implement Clinical Software for implementing a rapid response Public Health Safety and Treatment DATA Management System for NOVEL Coronavirus-Convid 19 Pandemic. The simulated diagnosis results were obtained from preclinical model simulation algorithms in MATLAB software environment to assess CONVID-19 antibodies vaccine treatment using regression fits from neural network modeling in a clinical software decision support system developed. The two important landmarks of this research study relates to specific anatomical physiologic parameters for post-treatment clinical trials to be conducted for CONVID-19 infection pathology factorials and combinations. Thus the knowledge gain from the initial preclinical simulation research study points to new leads in finding a cure for the CONVID-19 using concept of genome immune antibody vaccine to bridge knowledge gap of the DNA genomic proteome sequence to management CONVID-19 human immune deficiency virus. The most significant observation from preclinical simulated data shows that even though virus infection has been postulated by the experts may have nocure the findings from this work showed that the CONVID-19 exhibited disease symptoms within life cycle, patient immune system is the ultimate providing the change maker catalysts agent to make antibodiesto cure the patient to a functional state within 21 days of the virus lifecycle.

Keywords: Genome, CONVD-19, Antibodies, Immunity, Viral infection, Clinical data, Preclinical trials

1. INTRODUCTION

The CONVID-19 biomarker in human immune pathology targets single-nucleotide polymorphisms of the infected patient physiology observed through preclinical disease symptoms manifestation. According to the report by [1, 2], a novel respiratory virus that originated in Wuhan China has spread to more than one hundred (100) countries combined located in Asia, Europe, North America, Africa and Middle East. More than Three million Five hundred thousand (49, 000,000) cases of people infected with this virus as of October 28, 2020 have been reported and new cases are increasing exponentially making the world health organization declaring the Novel Coronavirus –COVID-19 as a pandemic. The United States alone has reported more than Eight Million Five hundred thousand (8, 500,000) cases with 23,100 deaths according to a recent CNN cable news report and New York Times [3]. The virus is effective in transmitting through communities through airborne or physical contact with experts confirming doubling every six days according a published model [4, 5]. Much remains unknown about the virus, however many people may have very mild or asymptomatic infections [6].

The economic impact of the virus to the states and countries translates directly and indirectly into trillions of USD as recently reported by the US [7]. Coronavirus are named for the spikes they produced that protrude from their surfaces resembling a crown or sun's corona. COVID-19 infects human and animals and cause illness of respiratory tracts although its impact is less pronounced in human and animals of certain nutritional diets and progressed using biotechnology informatics techniques [8].The subject of genome organization and bioinformatics evolution of the Human DNA and virus nNRAfalls under a subclass of DNA genomics proteome bioinformaticsscience class [9, 10, 11, 12]. Coronavirus are enveloped non-segmented positive sensed RNA viruses belonging to the family coronaviridae and the other nidovirales and broadly distributed in human and other mammals [13]. Coronavirus is a single RNA protein surrounded by lipids consisting of fatty acids. Virus may have played a big role in the evolution of humans DNA of terrestrial intelligent beings.

The mutated DNA genomics and bioinformatics codes could have been passed by parent Homo sapiens to its progenies generations through procreation to make species of human DNA Proteome immune system that forms antibodies to containthe virus epidemiology. The first Homosapiensgenes could have been mutatedby infection of nNRA single virus strain in the homosapiens to form complex intelligent DNA genomics code. The generation of DNA gene coding by mutation is an important step in understanding virus epidemiology in a human gene host. The virus primaryfunction may not be to infect a patient but to mutate the DNA donor complex genomes codes populated to its progenies. The Coronavirus nRNA strains change the DNA nucleotides code of its hosts and those who survive and passed thevirus generic code integrated intoDNA embedded platforms.

This is made possible through a viral bioinformatics amino acids protein sequence code in antibodies produced in complex DNA chromosome genomes to securefuture virus cells descendent by procreation in its donorhost DNA immune system. This way, the DNA nature recreates generic Genome DNA Proteome codes (Thynine, Adenine,Cysteine, Guanine, Uracil) (tacgu), RNA)code classifications to mirror (1, 2, 3, 4, 5, 6) numeric codes converted into binary bit codes (0,1, 10, 11,100, 111). The DNA genomics organizations and antibodies bioinformatics evolutions in clinical disorders and reorders of the existing DNA strain as DNA gene resists changing its codes from the invading rNA virus (a terrestrial genetic intelligent single cell of another biological organism in pseudo states from planetary organisms) thriving to mutates its bioinformatics code architectures. Thus a CONVID -19 virus nRNA penetrate the donor DNA genomics and bioinformatics code of its donor and mutates donor to propagate gain dominance of its host.DNA complex Genome architectures and bioinformatics coding involves including a viral codeto mutate DNA trillion genesinto another generic hosts.

This may be how the DNA genomicssequence its antibodies bioinformatics complex code architecture when it has it has come of age. Major theory gaps in knowledge of origin, epidemiology, duration of human transmission and clinical prognosis of CONVID-19 viral infection pandemic has necessitated further research in the finding the appropriate clinical technology application to simulate antibodies to limit virus population, propagation, transmission and epidemiology on the DNA gene framework. The modeling of DNA genomics sequence and antibodies bioinformatics evolution of possible CONVID-19 for infected patients were simulated in MATLAB software environment. A fit of clinical expectations of viral infected immune antibodies index for four patient cases selected from clinical data set was simulated in MATLAB software environment to model infection-immune mapping and antibodies evolution in clinical decision support software. Patient 0 which is primary carrier of CONVID-19 virus exhibited the highest disease state of 47% followed by Patient 1.

Patient 0 simulated diseased states is 47%, patient 1, 45%, Patient 2, 6% and Patient 3, 3%. Diseased states were modeled as Bayesian probability distribution variable between 0 and 100% (0 and 1) subject to a set of diseased conditions from the degree of symptoms experienced by each patient. Patient 3 shows the least disease states. Even though patient 1 shows highest antibodies index of 43% compared to patient 0, which showed 39%, the disease state in patient 0 is highest at 47% compared to 45% of patient 1 in comparison within 3-6 days. Patient 0 shows reduced disease load of (0.401072) after 18-21 days from a high value setting from (0.776167) within 6-9 days because of increase treatment index (0.581956) from (0.358707). The immune index for Patient 1 and Patient 2 reaches an all-time high of 0.9910997 and 0.978334. The preclinical simulation results data showed reduction of CONVID-19 infection for Patient 0, the index case reduced to 34% within 18-21 days. Thus Patient 0 shows immunity-infection indexes reduction from 62% within the 3-6 days, and to less 38% within 18-21 days

2. PRE CLINICAL MODELLING AND METHODS

A comprehensive description of the eligibility criteria used to select trial clinical data for preclinical MATLAB simulation relies on the submissions by clinicians in charge of the data to provide unbiased data. The data were subjected to further test for accuracy using appropriate software regression analysis testing in MATLAB software environment under certain published protocols and guidelines. This preclinical study was conducted following the principles as applicable to UCSF Policies and procedures and all applicable federal and state laws and regulations including 45 CFR 46, and the HIPAA Privacy Rule. The study also incorporates sound scientific basis and tested algorithms to develop clinical software to model antibodies evolution for CONVID-19 viral infection in the human system which assumes thatcorona virus falls under the class of acquired human immune deficiency virus that affects the immune system.

The trial candidates for clinical guidelines development include an initial software prognosis on clinical data. There are acquired and non-acquired human deficiency virus and where immune system is properly functional, disease symptoms are not possible. However, all acquired viruses like HIV virus, Corona-virus;CONVID-19 virus affects the biomarker in the immune DNA structure important in preventing disease symptoms to manifest. None acquired immune viruses are those viruses that degenerate within the human host such as Cancer causing virus, Alzheimer's neurodegenerative disease virus. These thoughts for virus clinical manifestation haveled to development ofthe new science to determine the epidemiology and genomics evolution pathological pathway of corona virus [9]. Novel beta Corona virus CONVID-19 like Human immunodeficiency virus type 1 (HIV-1) encodes three enzymes which are required for viral replication: reverse transcriptase, protease, and integrase (IN) [10]. The environment in which a preclinical randomized trial is conducted in the clinical software system may differ considerably from the setting in which the trials are used and applied in a controlled clinical trial with real time and life subjects. Clinical Data used were subjected to statistical procedure to determine best fit on covariance. Although drugs targeting reverse transcriptase and protease could be used and may show some effectiveness, however combinations of toxicity and resistant strains have limited usefulness.

Therefore there is an ever increasing drive to discover new drugstargets against alternative sites in the viral life cycle.IN has thus emerged as an attractive target for anti-HIV therapy, both because it is necessary for stable infection and because known functional analogues are lacking in the human host [10]. Integrase inserts a double stranded DNA copy of the viral RNA genome into the chromosomes of an infected cell through two separate reactions; in the "3¢-processing" step, IN removes two nucleotides from each 3¢ end of viral cDNA, while in the "strand transfer" reaction, the two newly processed 3¢-viral DNA ends are inserted into the host cell DNA. For the integration reaction, no source of energy (e.g. no ATP) is needed and only divalent cations such as Mn^{2+} or Mg^{2+} are required for the catalytic activity. Although a wide variety of compounds have been reported as IN inhibitors, 3 drugs active against this enzyme have not as yet been approved by the FDA.

One of the major leads in the development of anti-HIV-1 IN drugs is represented by â-diketo acids4 (DKAs) and their derivatives3 which have been shown to selectively inhibit the strand transfer step by sequestering the divalent cation bound in the active site of the enzyme,5 and to block HIV-1 replication in infected cells. Considering that the only two IN inhibitors so far in clinical trials, 3 Merck's L-870,810 and ShionogiGlaxo-SmithKline's S-1360 (Figure 1, 1 and 19), belong to the â-diketo-like acid family, recently constructed a three-dimensional pharmacophores model for the binding of DKAs to the enzyme. Using this model as a query for virtual screening, several compounds that contain the specified 3D patterns of chemical functions have been found. Biological testing showed that strategy was successful in searching for new structural leads asHIV-1 IN inhibitors [11].Several research investigations have been extended to the development of 3D QSAR models for the same family of IN inhibitors. Even if several HIV IN pharmacophores has already been published, 7-12; to the best of knowledge no attempts to build quantitative models for DKA-like strand-transfer-selective inhibitors have as yet been reported.

The only two previously reported pharmacophores for DKA-like derivatives are "qualitative" hypotheses 6, 12 (generated without the use of activity data), whereas also more sophisticated "quantitative predictive" model with biological activities were taken into account has been developed. The novel potential class of â-diketo acid containing inhibitors of IN, and the evaluation of the IN inhibitory activity confirmed the strength of this rational approach Over a series researchers and publications have previously reported detailed studies of a class of 4, 5, 6, 7-tetrahydro-5- methylimidazo [4, 5, 1-jk] benzodiazepin-2-(1H)-one com- pounds (TIBO), 1. These compounds possess significant inhibitory activity against the human immunodeficiency virus type 1 (HIV-1), the causative infection of acquired immune deficiency syndrome (AIDS).1±5 other groups have explored the possibility of maintaining anti-HIV activity with modified versions of the tricycle 1. These modifications ranged from extensive examination of the structural requirements of 1, which resulted in the discovery of the comparatively active PETT series of compounds, 6 to simpler modifications, such as solely opening either the5- or 7- member ring of 1, i.e. 2 and 3, respectively **[12]**.

A group that prepared compounds akin to 2 reported that no anti-HIV activity was detected for their truncated versions of 1-7 From the independent study examining disruption of the 7-membered ring of 1, it was reported that modest anti-HIV activity is maintained (IC50=4±30 mM) after modification. The findings of researchers related to anti-HIV structure activity relationships (SAR) of 5-membered ring truncated analogues of 1, as depicted by generic bicyclic structure 4, as well as the associated synthetic preparations [13].Cell proliferation plays an essential role in immune responses. Initial successful efforts with chemical immunosuppressant were used, utilizing nucleoside antagonist 6-mercaptopurine (or its imidazole, analog azathioprine), which have its side effects (bone marrow suppressions with emergence of serious infections) [14, 15]. Other variety of chemical antiproliferative agents over the years has been developed without overcoming the hazards of azathioprine.

There are documented nutritional and vitamin deficiencies in early stages of HIV infection, and malnutrition associated with increased risk of HIV disease progression [23, 24, 26]. There have also been long standing interest in complementary and alternative medicine (CAM), approaches to managing HIV infection and various conditions associated with HIV, which could be applied to CONVID-19 treatment. The CAMs commonly used for people living with HIV are not drugs, herbs or other pharmacology agents, but rather things like meditation, energy building and acupuncture could offer tremendous healing therapy [13, 19]. So far technology forHIV and Novel COVID related respiratory virus control and management has been to suppress the HIV- virus from proliferation [12]. U.S. published patent application US 2005/0080260 A1, to Mills et al., which teaches the synthesis of a chemical compound having the formula A-B-C that may serve for applications such as drug delivery where A is a chemiluminescent, moiety, B is a photochromic moiety, and C is a biologically active moiety where A-B-C may serve as a prodrug. Novel synthetic methods of the method to form the prodrug comprised the steps of (1) forming abenzophenone, (2) forming a diaryl ethylene, (3) attaching a phthalimide moiety to at least one of the aryl groups of the ethylene to form a phthalimide-ethylene conjugate, (4) condensing two ethylene-phthalimide conjugates to form a phthalimide-pentadiene conjugate, (5) converting the phthalimide to the phthalhydrazide by reaction with hydrazine to form a carrier compound according to the present invention, and (6) reacting the carrier compound with an nucleophilic moiety of the drug to form the corresponding prodrug.

2.1 Modeling of Covid-19 (nNRA) and (DNA) Antibodies Bioinformatics Evolution-Human genome organization and antibodies bioinformatics protein sequence code architecture mutated by Coronavirus CONVID-19 infection show relative reading frames of individual genes. Both the tat and rev genes are encoded in two axons, in different reading frames as indicated by identical shaded areas. The different genes and their functions are: gag, core proteins necessary for capsid formation pol, enzymes required for virus replication; env; external glycoproteins; vif and nef, probable regulatory proteins; tat and rev, regulatory proteins essential for virus replication; vpr, unknown function. Kb, kilobases [16] (source: Gallo et.al. 1988). The CONVID-19 virus traced to Wuhan market in China could be curtailed by washing hands with alcohol or methanol spirit to break down the lipids, and disintegrates the virus, which is why washing hands with gels or alcohol could help reduce spread. Virus is a single RNA protein surrounded by Lipids is a very basic gene proteome structure compared to the 12 trillion DNA genomes formed in the human system. The virology science is not to assume the virus as a disease that harm the patient but a single nRNA mutating the complex human DNA code to exert its dominance in future progenies gene hosts that survive in a complex human DNA immune antibody system. The assumptions is that virus infection recodes the complex DNA genomes of its host to mutate it into its own DNA genomes proteome structure, a phenomenon refer to as DNA genome reorganization and Bioinformatics evolution to form another generic intelligent terrestrial species using the human DNA as host pods. There is little knowledge of the origin of virus, but it could be terrestrial organism trying to find its own generic identity but multiplying its own RNA in our complex human DNA system by mutation. The final product of this mutation process is another generic population.

The study of Huang et.al DOI: http// doi.org/10.1016/so140-6736 (20) 30 83-5 reports epidemiological, clinical, laboratory and radiological characteristics, treatment and outcomes of patients confirmed to have 2019-n-CONVID infection and compare clinical features between intensive care unit (ICU) and non-ICU patients. A novel coronavirus which was named 2019-nCONVID was isolated from lower respiratory tract specimen and a diagnostic test for this virus was developed soon after [3] of 59 cases, 41 Patients were confirmed to be infected 2019-nCONVID [3]. The presence of 2019-nCONVID in respiratory specimens was detected by next generation sequencing on real time RT-PCR methods. The primers and probe targets to envelope gene of CONVID-19 were used and the sequences were as follows:

Forward primer 5'-
ACTTCTTTTTCTTGCTTTCGTGGT-3';
Reverse primer 5'-
GCAGCAGTACGCACACACACAATC-3' and
The probe 5' CY5-
CTAGTTACACTAGCCATCCTTACTGC-3' BHQ1.

The protein sequence in DNA genomics immune proteomics databases search engines is inbuilt inhuman system retrieval protein identification protocols (PIR) sequential processprotocols. The PIR can identify and forcefully transport the novel CONVID -19 viral RNA cells in corona tract hosts' pods to bind to cloned immune system CONVID-19 T-gene that effectively destroy the viral infection. Further clinical studies suggest bio organics nutritional supplements bicarbonates salt solution enriched with fructose and glucose cofactors could hold prospects for development of prodrug antibodies for corona virus epidemic. Conditions for the amplification were 50°C for 15min, 95°C for 3 min, followed by 45 cycles of 95°C for 15s and 60°C for 30s [6,7]. The initial investigations for the sick patients in Wuhan, China that contracted the virus included a complete blood count, coagulation profile, and serum biochemical test (including renal and liver function, creatine kinase, lactate dehydrogenase and electrolytes). Respiratory specimens including nasal and pharyngeal swabs, bronchoalveolar lavage fluid, sputum, or bronchial aspirates were tested for common virus including influenza, respiratory syncytial virus, adenovirus, Para influenza virus, SARs-CONVID and MERS-CONVID-19 using real time RT-PCR approved assays by the China Food and Drug Administration [6, 7].

Deoxyribonucleic acid (DNA) is the chemical compound that contains the instructions needed to develop and direct the activities of nearly all living organisms. DNA molecules are made of two twisting, paired strands, often referred to as a double helix Each DNA strand is made of four chemical units, called nucleotide bases, which comprise the genetic "alphabet." The bases are adenine (A), thymine (T), guanine (G), and cytosine (C). Bases on opposite strands pair specifically: an A always pairs with a T; a C always pairs with a G. The order of the As, Ts, Cs and Gs determines the meaning of the information encoded in that part of the DNA molecule just as the order of letters determines the meaning of a word [8]. An organism's complete set of DNA is called its genome. Virtually every single cell in the body contains a complete copy of the approximately 3 billion DNA base pairs, or letters, that make up the human genome. Therefore with its four-letter language, DNA contains the information needed to build the entire human body. A gene traditionally refers to the unit of DNA that carries the instructions for making a specific protein or set of proteins. Thus each of the estimated 20,000 to 25,000 genes in the human genome codes for an average of three proteins. Located on 23 pairs of chromosomes packed into the nucleus of a human cell, genes direct the production of proteins with the assistance of enzymes and messenger molecules. Specifically, an enzyme copies the information in a gene's DNA into a molecule called messenger ribonucleic acid (mRNA). The mRNA travels out of the nucleus and into the cell's cytoplasm, where the mRNA is read by a tiny molecular machine called a ribosome, and the information is used to link together small molecules called amino acids in the right order to form a specific protein.Proteins make up body structures like organs and tissue, as well as control chemical reactions and carry signals between cells.

Thus when a cell's DNA is mutated, an abnormal protein may be produced, which can disrupt the body's usual processes and lead to a disease such as cancer.The novel coronavirus (CONVID-19) are classified as lentiviruses based on their morphology genomic organization, percentage of nucleic acid identity to RNA and cross reactivity of antibodies to proteins of other lentivirus. Matured lentivirus particles contain an eccentric bar shaped core that is form after the virus form from the cell membrane. The enveloped glycoprotein of the novel

Coronavirus has two properties in common, the enlarge spikes after it buds and the many number of glycosylation sites by [13, 14, 15]. The tat and rev genes code for proteins with regulatory functions that are essential for virus replication and probablyact at both transcriptional and translational levels. The coding sequences for both of these genes are novel for retroviruses and are composed of multiple exons [16, 17] (Arya et.al 1985; sodroski et.al 1985). The function of the vif, nef and vpr gene products are not known. All of the proteins, including those regulatory and those unknown functions are recognized by serum from a Coronavirus (CONVID-19) infected persons

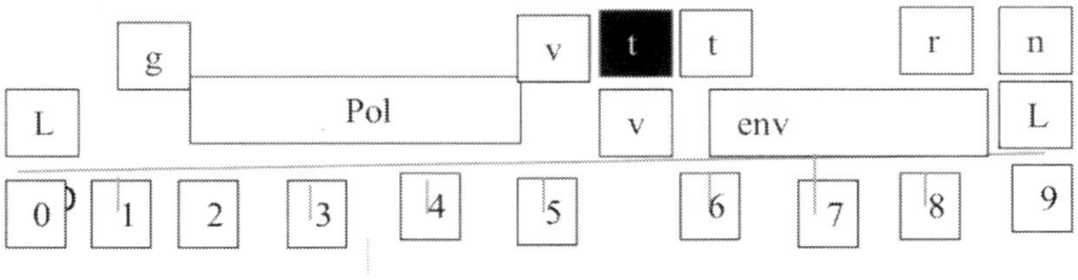

Figure 1: Genome organization of Coronavirus showing relative reading frames of individual genes.

Both the tat and rev genes are encoded in two exons in different reading frames as indicated by identical shaded areas. The different genes and their functions are: gag, core proteins necessary for capsid formation; pol, enzymes required for virus replication; env, external glycoproteins; vif and nef, probably regulatory proteins; tat and rev, regulatory proteins essential for virus replication; vpr, unknown function. Human beings are carbon based organisms and mammals. However the DNA genomics and mRAN proteomes and regulatory functions of immune system including brain serology and viral epidemiology affect the protein based structures of the human system making the case that does with high more than 80% protein diet as those in China would likely suffer greater clinical pandemic than those in Nigeria where 80% of diet is rich are carbohydrates. To understand the underlying bioorganic chemistry of the corona-virus in the human system we discuss the basic chemistry underlying Convid-19 viral infection serology in the human system. The 2019-nCoV infection caused clusters of severe respiratory illness similar to severe acute respiratory syndrome coronavirus as was associated with ICU admission in Wuhan China as reported by Huang et.al DOI: http// doi.org/10.1016/so140-6736 (20) 30 83-5. The major gap in their knowledge of the theoretical basis for the origin, epidemiology, and duration of human transmission and clinical spectrum of disease forms the basis for this study.

2.2 Proteomics Evolution, Genomics Serology and Disease Epidemiology-The Proteomes genomes are set of proteins of the DNA gene organism. Proteomics combines the census, distribution, interactions, dynamics and expression patterns of protein within living systems [17]. DNA microarrays can be used to determine expression patterns of different proteins by detections of mRNA's or for genotyping by detection of different variant gene sequences including but not limited to single nucleotide polymorphisms (SNps). The application of DNA microarrays involved identifying genetic individuality in tissues or organism in human, investigating cellular states and processes patterns of expression that change with cellular state or growth conditions, diagnosis of disease which involves the testing of mutations in proteomics , genetic warning signs, whereas some disease are not identified or irrevocable, drug selection detection of genetic factors that governs response to drugs, specialized diagnosis of disease and target selection and drug design. The algorithm for proteomics evolution, genomics serology and disease epidemiology engineering is presented below:

(1) **Diagnosis of Disease**: Testing for the Presence of mutations can confirm the diagnosis of a suspected genetic disease

(2) **Genetic Warning Signs**: Some disease are not determined entirely and irrevocably by genotype, but the probability of their development is correlated with genes or their clinical expression patterns

(3) **Drug Selection**: Detection of genetic factors that govern responses to drugs, that in some cases render treatment ineffective and in others cause unusual serious adverse reactions

(4) **Specialized Diagnosis of Disease**: Knowing the exact type of disease is important for prognosis and for selecting optimal treatment

(5) **Target Selection in Drug Design**: Proteins showing enhanced transcription in particular diseased states might be candidates for attempts at pharmacological intervention

(6) **Pathogen Resistance**: Comparisons of genotypes or expressions patterns between bacterial strains susceptible and resistant to antibiotic point to point the proteins involved in the mechanism resistance

(7) **Temporal Variations in Protein Expression**: Permits timing the course of responses to pathogen or viral infection , responses to environmental change, changes during the cell cycle

(8) **Mass Spectrometry**: Mass spectrometry is a physical technique that characterized molecules by measurement of their masses of their ions. Applications to Proteomics include

  (a) Rapid Identification of the Components of a complex mixture of proteins

  (b) Sequencing of Proteins and Nucleic Acid

  (c) Analysis of post –translational modifications, or substitutes related to an expected sequence

  (d) Measuring extents of hydrogen –deuterium exchange to reveal the solvent exposure of individual sites. This provides information about static conformation dynamics-including folding and interactions

(9) **System Biology Integration**: System biology integration has two aspects a) The study of patterns within cell or an organisms, patterns of protein-protein, and protein-nuclei acid interactions, patterns of metabolic pathways and control cascades and patterns of protein expressions. Patterns have both static and dynamic aspects 1) Identification of pairs of proteins that bind to each other, and the assembly of pairwise interactions of proteins that bind each other and assembly of pairwise interactions into a network of enzymes of the flow of information down a control cascade is dynamic patterns b) The other aspects of integration is the comparison of occurrence, activities and interaction of genes and proteins across different species. High throughput methods of genomics and proteomics provide data about sequences, expression patterns and interactions. From Genome sequences we can infer the amino acid sequences and sequences of an organism complement of proteins. Proteomics tells us how expression patterns of proteins vary, within the organism, how they change or evolved during development in responses to changes in the environmental conditions and how they cooperate with one another. System biology takes these data and integrates them to extend in both space and time.

(10)**Clinical Implications:**

(i) Gene Diagnosis- There is a general consensus that the sequencing of the human and other genome will lead to improvement in health care by 1) Genomics Diagnosis of disease and risks associated: DNA sequencing can detect the absence of a particular gene or a mutation. Identification of specific gene sequences associated with diseases will permit fast and reliable diagnosis of conditions (a) when a patient presents with symptoms (b) In advance of appearance of symptoms as in tests for inherited late onset conditions (c) For in utero diagnosis of potential abnormalities such as cystic fibrosis (d) for genetic counseling. In many cases our genes do not irrevocably condemn us to contract a disease but raise the probability that we will. An example of a risk factor detectable at the genetic level involves $\alpha_1 -$ antitrypsin (342 Glu-Lys) express only a dysfunctional protein. Liver diseases are caused by accumulation of polymeric form of $\alpha_1 -$antitrypsin in hepatocytes where it is synthesized. The relationship between genotype and disease risk is much difficult to pin down. Some diseases such as asthma and CONVID-19-Convid-19 (19 CONVID) depend on the interactions of many genes as well as environmental factors, so that a mutation elsewhere may alter its level of expression or distribution among tissues. Such abnormalities must be detected by measurements of protein activity. Also analysis of protein expression patterns is also important way to measure response to treatment.

(ii) Genetics of Therapy response: Customized Treatment Pharmacogenomics-Different people differ in ability to metabolize drugs; different patients with the same conditions may require different dosages. Sequence analysis permits selecting drugs and dosages optimal for individual patients, a fast growing field called pharmacogenomics resulting in billions of dollars savings in health care costs and could be applied to 19-CONVID

(iii) IdentificationofDrugTargets: A target is a protein the function which can be selectively modified by interaction by a drug to affect the symptoms underlying causes of diseases. Identification of a target provides the focus for subsequent steps in the drug design process. Among drugs in use the targets are about half are receptors based, about a quarter are enzymes based and about a quarter are hormones. Approximately 7% are unknown targets, but none of the drugs developed so far are immune based which is the focus of our research, how we have synthesized nutritional supplements in bioorganic oral solutions of bicarbonates and ionized water salts carefully selected to interact with the gene targets in the immune system to stabilized the biomarkers or improved the immune biomarker or restore the immune biomarkers to target receptors, enzymes, hormones and the pathogens to mutate or reprogrammed the gene therapy to an appropriate targets in space and time.

(iv) Gene Therapy: If a gene is missing or defective it can replaced or at least supplied its deficiency using nutritional supplements in bioorganic oral solutions. Direct supply of proteins is possible for many diseases of which insulin replacement for diabetes. Gene transfer has succeeded in animals for production of human proteins. Gene replacement therapy is blocking genes is called antisense therapy. Binding to endogenous DNA can interfere with transcription; binding to mRNA can interfere with translation. Antisense therapy has shown some efficacy against cytomegalovirus and Crohn' disease

(v) Prodrug nutritional supplements from Banana bioorganic mixture is an immune booster could provide pro drugsto target at the CONVIDID-19 virus epidemiology in the Human System. Other drugs which attempted to improve resistance of immunity of the Gene -cells of the (CONVID19)-immune Gene cell to the virus have left other side effects. Antiviral drugs used for CONVID-19- Genomics Treatment and Management have failed so far because the CONVID19 has ability to recreate into another strain of virus to render proteomes cells impotent. Banana Prodrug boost the immune system to create a antibodies in the biomarker in the Corona tract to fight the virus in the human tract by limiting the CONVID-19 stearic and measomeric inductive resonance that mutate Proteome-DNA system. The CONVID19 is not the disease that eventually destroys the patient; rather it is other minor or major ailment, which invades the human immune system and renders it incapacitated to defend itself from any diseases. The CONVID-19creates a situation of AIDS-acquired immune deficiency.

2.3 Methodology for Clinical Simulations Solution derived from Data -There were no human subjects used for our preclinical model simulation studies and ethical approvals were resolved by removing the names of patient's clinical data hematology and serology in the final data presented in Appendix A1. Clinical data used in developing scientific modelbasis analysis were derived from Lagos hospital in Nigeria and study of Wuhan China as reported Huang et.al DOI: http// doi.org/10.1016/so140-6736 (20) 30 83-5 for patients admitted into the hospital confirmed to have N-CONVID-19 infection. Most of the infected patients were men 30(73%), including diabetes (eight (20%), hypertension six (15%). Median age was 49.0 years (IQR 41.0-58.0), 27(66%) of 41 patients has been exposed to the seafood market. The family cluster exhibited common symptoms at onset of illness were fever (40(98%) of 41 patients), cough (31(76%) and myalgia or fatigue (18 [44%]. The less common symptoms were sputum production (11 (28%) of 39), headache (three [8%] of 38], hemoptysis (two [5%] of 39] and diarrhea (one (3%) of 38). Dyspnoea developed in 22 (55%) of 40 patients (median time from illness onset to dyspnea 8.0 days [IQR 5.0 -13.0]. 26 (63%) of 41 patients had lymphopenia. All patients had pneumonia with abnormal findings on chest CT. Complications included acute respiratory disease syndrome (12 [29%]), RNAaemia (six (15%)), acute cardiac injury (five (12%) and secondary infection (four (10%)). 13 (32%) patients were admitted to an ICU and six (15%) died. Compared with non-ICU patients, ICU patients had higher plasma levels IL2, IL7, IL10, GSCF, IP10, MCPI, MIPIA and TNF. The conclusion from work of Huang et.al coauthors are that 2019-nCONVID infection caused clusters of severe respiratory syndrome coronavirus and was associated with ICU admission high mortality rate. There is consensus that the sequencing of the human and other genomes that could lead to improvements in the health of Human Carrier of Corona Virus (CONVID19). Corona Virus (CONVID19) Containment applications described in the proposed clinical guidelines includes: According to Abhulimen (2016), [20] model for the HIV; CONVID-19 virus mirror the pathology of the virusproteome sequence modeling as follows:

1. Diagnosis of Disease and Disease risks: DNA sequencing is recommended as this can detect absence of a particular gene and mutation. Decision Support System (DSS) for disease diagnosis could assist in DNA sequencing. Identification of specific gene associated with Corona Virus (CONVID19) disease will permit fast and reliable diagnosis of condition for mapping of the prodrug compound that would produce the right antibodies within the Human Immune System to retroactively retract the virus RNA for the Corona Virus (CONVID19) antibodies production within the patient

2. When a corona virus (CONVID19) patient presents symptoms in the mucous of the ear, nose and throat, and lungs, (Gene DNA-Corona Tract)-Proteome Gene work, heart cardiograph and all tests for early screening and monitoring would provide basis for data feed for our software DSS to provide the correct DNA sequencing, for production of the right pro drug agent to stimulate a virtual immune system to produce the antibodies through the adaptive immune system on the innate system body that eventually produced the antibodies from the virus RNA thereby curing the patient. Further tests such as advanced DNA sequencing decision support system developed in our research to screen for the virus in infancy and matured phase of the virus light cycle.

3. Gene'smutations by a virus does not irrevocably condemn us to contract a disease but raise probability that we will do. Often the relationship between genotype and disease risk is much difficult to match. Some diseases such as Asthma depend on interactions of many genes and environmental factors. In some cases a gene may be all present and correct, but a mutation elsewhere may alter its level of expression and distribution among tissues observable through abnormality of protein activity. Analysis of protein expression patterns also an important way to measure response to treatment. Corona Virus (CONVID-19) disease presents such complex immune DNA sequencing pathways and mutations. The Corona Virus (CONVID-19) uses the host DNA to produce the human genome deficiency sequence that eventually leads to the death of the host. In providing antibodies our clinical approach reverses the trend by using the Corona Virus (CONVID19) RNA to produce the Antibodies for the cure of the infected host Patient.

4. Immune system genetics applied for the corona virus (CONVID19) clinical therapy. Path physiology and clinical manifestations. Different doses are required because of patient's different ability to metabolize drugs. Sequencing analysis permits selecting drugs and dosages optimal for individual patients, a fast growing field called pharmacogenomics, immune genome genetics sequencing is best done using our decision support system studio developed as part of our clinical guideline for containment of Corona Virus (CONVID19).

2.4 **Immune System Clinical Manifestations** -There are three immunology System [16]; The Body Innate I_1, Adaptive I_2 and a third that has not been researched, the Virtual Immune System I3 existing within the domain of unchartered clinical research space . The virtual immune system I_3 provides specific Bioinformatics instructions through the Adaptive Immune system T8 to the body to produce the Antibody specific to a viral alien agent. The Virtual Immune system is the power house and source of all immune responses and can be the site of virtual intelligence as it enlivens the other immune system, adaptive and the innate immune system.

B-Defense (Innate)

$$\boxed{V_1 = Vaccine\ Load}$$

T-Propagation (Adaptive)

V-Create (Virtual Power)

$$\boxed{E_1 = Viral\ Load}$$

Viral Disease is at Subatomic Level [20]

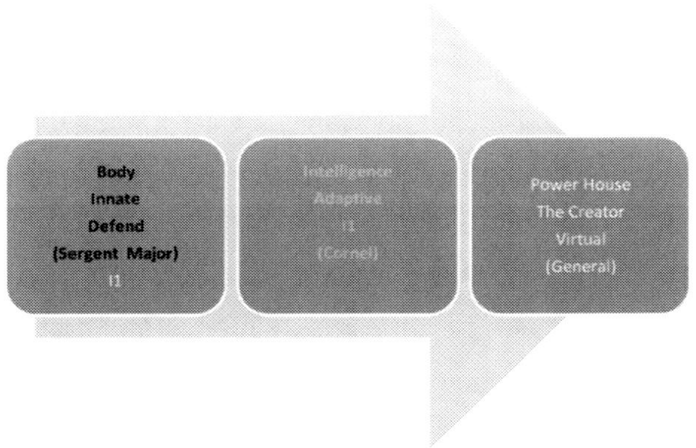

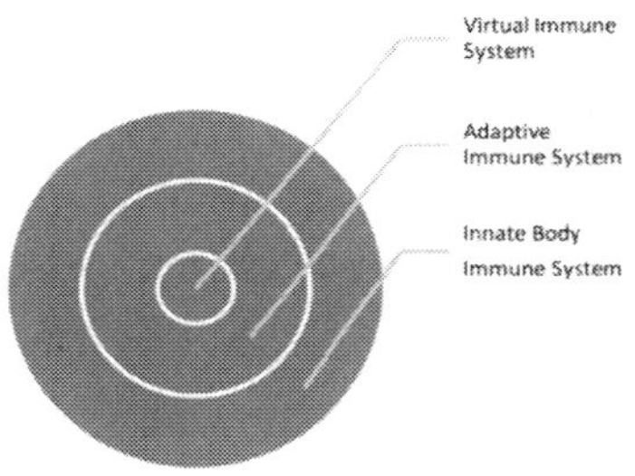

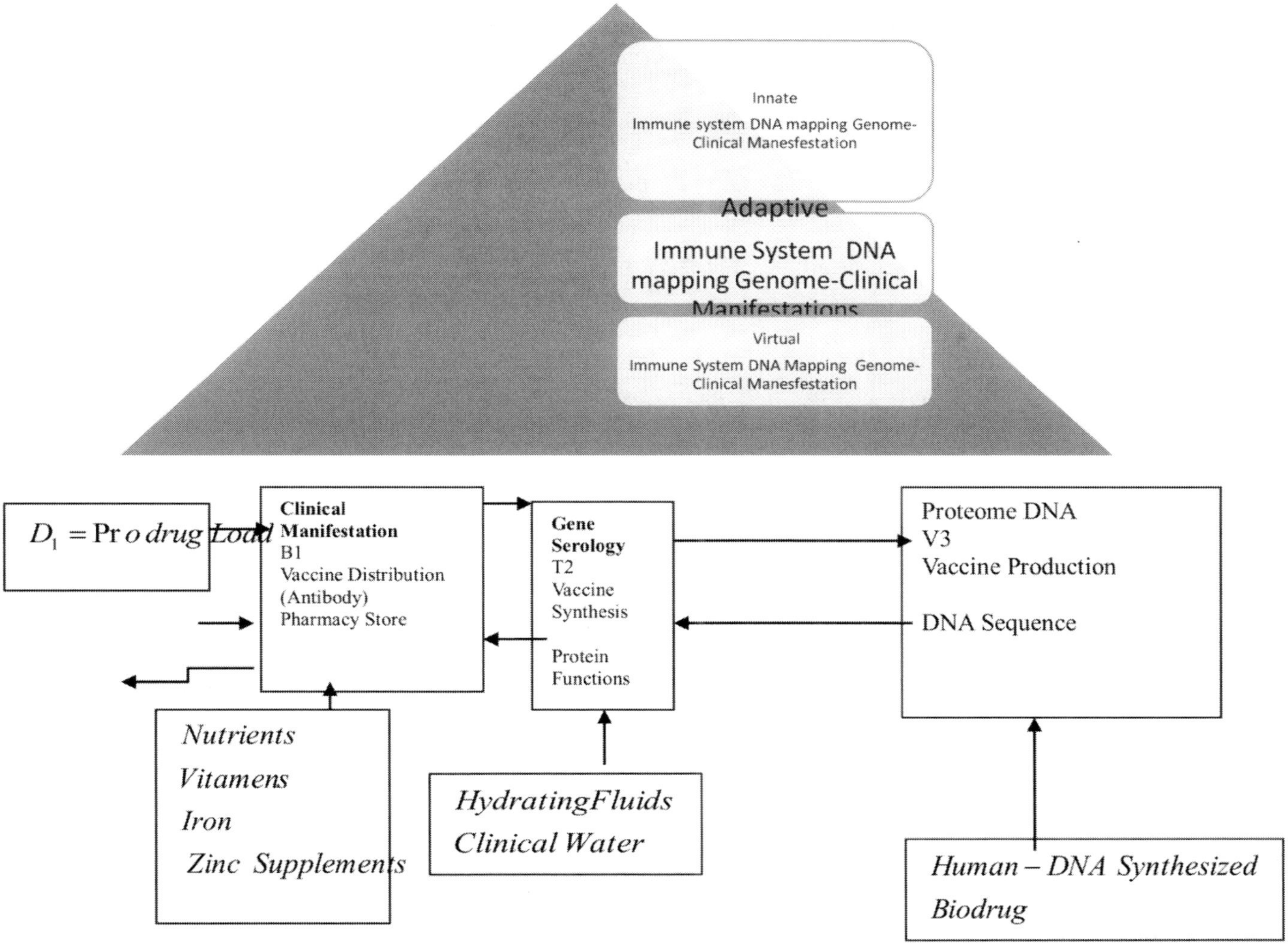

Figure 2: The DNA Immune System Genetics

2.5 CONVID-19 Epidemiology and DNA immune Clinical Genetics-The following step that describes CONVID-19 epidemiology and immune clinical genetics manifestation detection and containment is presented below:

1. **Identification Antibodies and Drug Targets:** A target is a protein which function can be selectively modified by interaction of a drug to affect the symptoms or underlying cause of a disease. Identification of a target provides the focus for subsequent steps in the drug and antibodies design process. Among drugs now in use, targets of about half are receptors, about a quarter enzymes, and about a quarter hormones. Growth in bacterial resistance to antibiotics is creating crises in disease control. Nutritional supplementsand antibiotic could genetically target the immune system to release the virtual power bioinformatics code sequence to produce of the antibodies within the host body to cure the virus.
2. **Gene Therapy.** If a gene is missing or defective we replace it or at least supply it product. Direct supply of proteins is possible for many diseases by insulin replacement for diabetes.

3. **Corona Virus (CONVID-19) Antibodies Production Premise**: A unified clinically tested antibodies is yet to be developed or reported for Corona Virus (CONVID-19) that is produced within a human host. Antibodies developed should be able to contain the Corona Virus (CONVID-19) within fourteen (14) days following our clinical guidelines. Conventional treatments of the symptoms using chloroquine, quinine, antibiotics, oral water hydrating a dehydrated patient keeping the patient in the positive mental state, diet selection supplies the necessary immune energy, nutrient vitamins and proteins keeps immune system stable for the production of the Corona Virus (CONVID19) Antibodies in the human system.
4. **The antibodies carrier is a cloned –CONVID-19** proteomes from immune non-pathological gene derived from bioorganic nutritional supplements with glucose and fructose co-factors and water bicarbonates salts solutions within the patients bodies from a clinical procedure and introduced in the patient (Gene DNA-Corona Tract)-Proteome Gene stream, thereby populating Corona Virus (CONVID-19) antibodiesusing Corona Virus **(CONVID-19)** RNA strain. Corona Virus (CONVID-19) propagation reaches the end of it's the natural lifecycle-propagation the death phase by re-engineering the RNA of the virus to continually produce antibodies for the cure. The antibodies agent reverses the Immune system responses to produce the cure of the virus. The infected patient index now becomes the patient antibodies reference for similar (Gene DNA-Corona Tract)-Proteome Gene work and genetic makeup. The clinical guideline for Corona Virus **(CONVID-19)**antibodies production and application is rigorous and complex DNA re-engineering processes, that it would practically be impossible to produce the Corona Virus (CONVID19) antibodies outside the patient hosts. Producing the antibodies outside the patient host is not recommended.
5. **The virtual DNA mapping of Corona Virus (CONVID-19)**antibodies produced in patient genomics serology and protease is propagated through continuous control application of doses of drug antibodies carrier into the coronary tract body every 2-3 days to produce the Genome Kinetic-clinical manifestations in the proteome genomics serology and using pharmacogenomics we can design an CONVID19 adaptive immune system. Nutritional Supplements in the Bioorganic and Oral solutions with fructose and glucose bicarbonates water solutions are not the antibodies but only assist to facilitate a cure using the patient DNA genomics and Proteome sequencing to produce the antibodies in the corona tract and stem cells marrow of bones. The patient immune system has the resources to provide DNA data templates capable of providing decisions related to antibodiesCONVID-19 within the immune genome to cure the Corona Virus (CONVID-19)-viral disease tissues. The patient is kept alive by reducing burden on immune system through supply of the nutritional supplements and antibodies. Our clinical guidelines incorporate treatment therapies that allow the mapping of biological clinically manufactured antibodies carrier to stimulate the patient virtual immune system to create specific Corona Virus (CONVID-19) antibodies. The Corona Virus (CONVID-19) antibodies that cures the patient is produced from the RNA tissue of Corona Virus (CONVID-19) using the genome DNA to re-engineer and recreate the DNA-RNA viral antibodies from the infected tissue cell thereby curing the patient of the virus.
6. **The corona virus (CONVID-19) DNA-RNA antibodies** propagate within the patient (Gene DNA-Corona Tract)-Proteome Gene work to eventually cure the patient of the virus. The raw material for the virtual-

adaptive-innate immune system antibody creates the Corona Virus (CONVID-19) antibodies from the RNA of the Corona Virus (CONVID-19). The virus is mutated to death as the antibodies produced cure of the Corona Virus (CONVID-19) in the infected patient. The Genetic DNA code in the immune system creates the Corona Virus (CONVID-19) antibodies (Patient Zero) from the affected Patient X, within the Patient's body. The antibodies carrier sends specific bioinformatics instructions to produce the specific antibodies that propagate to eventual cure the patient of the Corona Virus (CONVID-19).

2.6 Antibodies Virtual Immune System Kinetics-The virtual immune system kinetics is initiated by a prodrug

$$V_1 \overset{K_1}{\longleftrightarrow} V^{\oplus}_1 \quad B_o \overset{K_1}{\longleftrightarrow} B_O^{\oplus} \quad I_1 \overset{K_1}{\longleftrightarrow} I^{\oplus}_1$$

2 days

Corona Virus (CONVID19) ProdrugImmune System

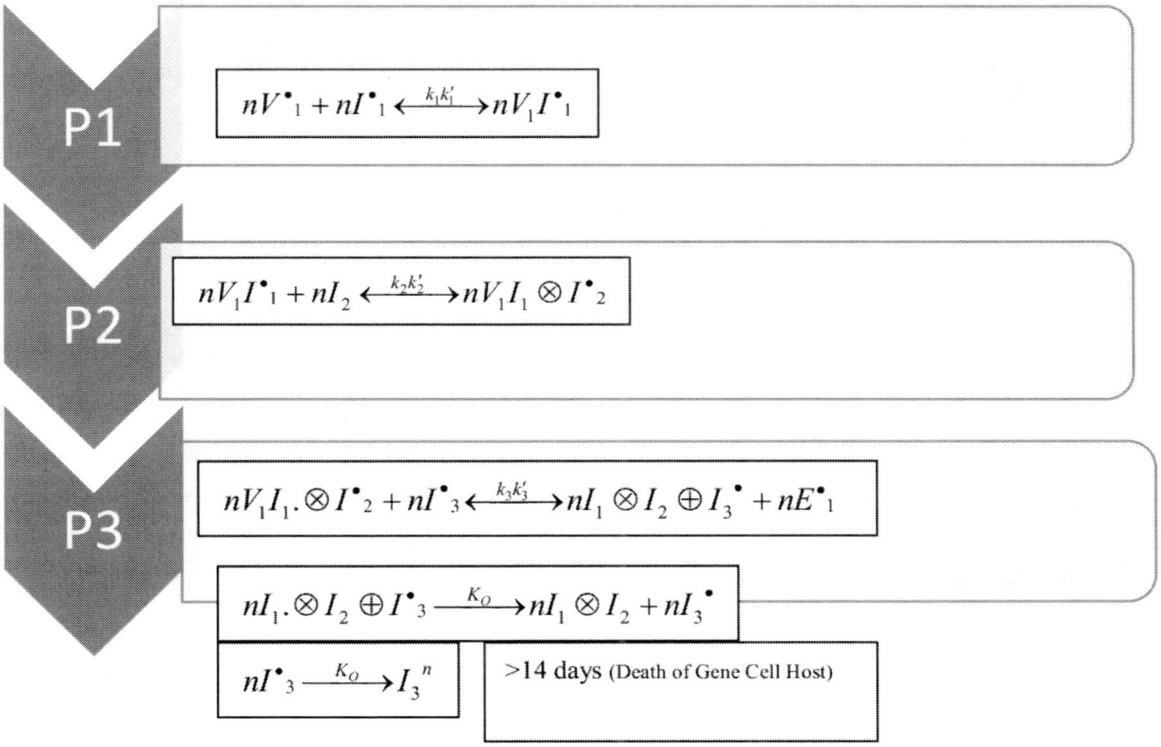

$$nE^{\bullet}_1 \overset{K_O}{\longrightarrow} E_.^{n}$$

>21 days
Virus propagated in Host

>21 days
Immune system stabilized at

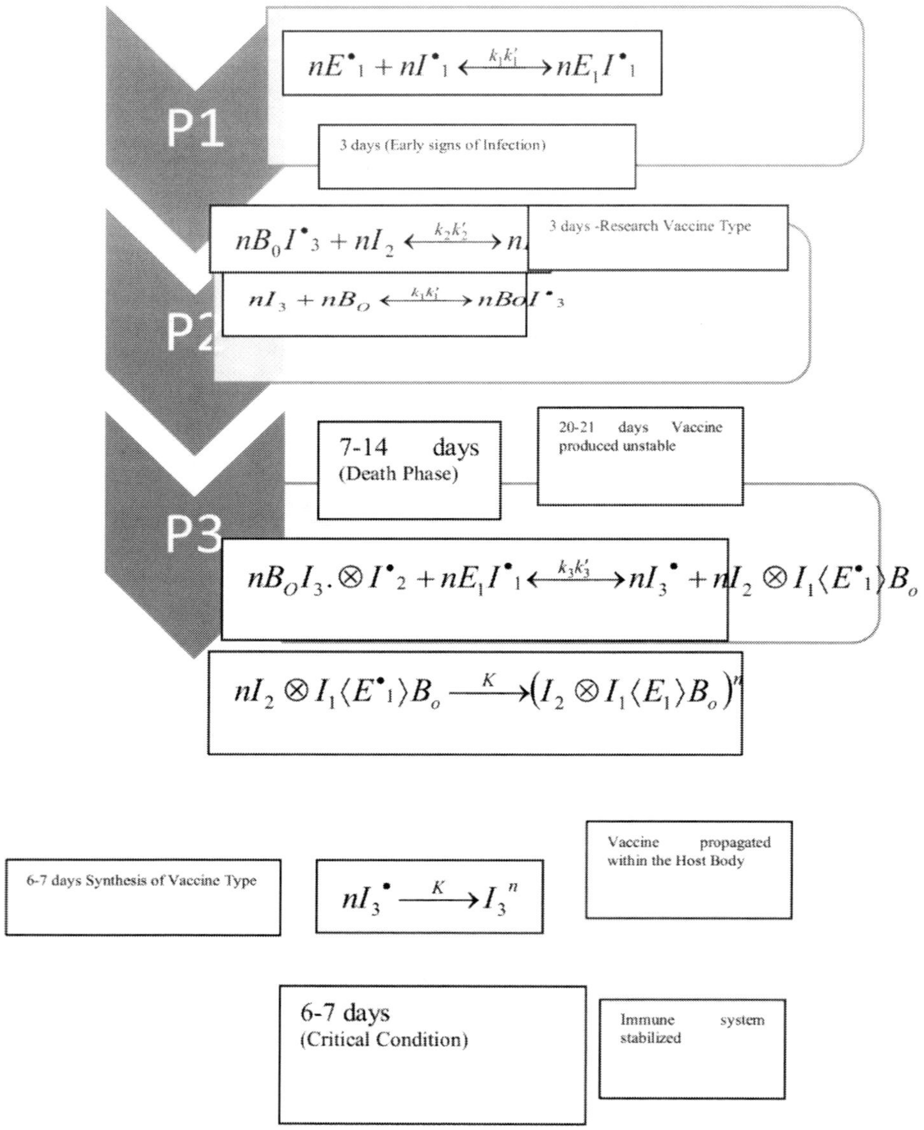

Figure 3.0: Corona Virus (CONVID19) Antibodies Carrier Kinetics

From stem cell research it is possible to recreate another identical CONVID- 19 antibodies clone cells. The CONVID- 19 gene clone is a natural gene antibody created using the RNA of the CONVID- 19 virus. The human system proteome gene cell infected by a CONVID-19- infection replicates to $(N-1)^{(N-1)}$ (CONVID-19)-+ and

CONVID-19-immune clone gene for a single viral strain of CONVID-19to populate to (CONVID-19)-+$(\mathrm{N}-1)^{(\mathrm{N}-1)}$. A-clone antibodies acts to shield Normal (CONVID19)-+ T-cells from propagation (Further practical research and clinical trials needs to be carried out to demonstrate this claim). The T-component in the antibodies in the white blood lymphatic to stimulates population of CONVID-19-immune clones' antibodies.

2.7MATLAB Simulation Model for Clinical Prognosis- Figure 4.0 shows the proposed DSS architecture model in a modified multilayer perception recurrent neural network (MLP RNN) according to Hagan *et al*, 1996; Haykin, 1999 training method according to (27, 28, 29) to model the COVID-19 and antibodies index preclinical studies. Fig.10 is the MLP RNN model trained as a feed-forward network by Levenberg-Marquardt (LM) algorithm [27, 28]. The input output data set Z^N of the system taken over NT period of time is available, that is: $Z^N\left(\left[u^I{}_N(t), y_N{}^I(t)\right] = 1,2.........N\right)$

[27, 28].Where $u^I{}_N(t)$ are input signals and $y^I{}_N(t)$ are the observed outputs. Part of the m inputs $u^{I1}{}_N(t)$ and $u^I{}_N(t)$ is applied to I_1 tap delay lines $(I_1 - TDL)$ memory. At time $(t+1)$, first n predicted outputs $y^{I1}{}_{Nm}(t)$ of $y^I{}_{Nm}(t+1)$ fed back to the input via another I_2tappeddelay line $(I_2 - TDL)$ memory. The contents of the two TDL memories are used to feed the input layer of the MLP to predict the state of the system. m and n corresponds to the number of past inputs and outputs used by the network for prediction; N is the total number of input output data pair (observations); N_m and N_n are the length of first m and n data fed to the MLP RNN (m and n are also of the orders of the line memories applied to the input and feedback signals respectively); I and I are therespective number of inputs and outputs vectors of the true system; I_1andI_2 are the respective number of inputs and outputs fed to the MLP RNN. Thus the output is ahead of the input by one time unit. The present and past values of the inputs $u^{Ii}{}_{Nm}(t), u^{Ii}{}_{Nm}(t-1)u^{Ii}{}_{Nm}(t-m)$ represent exogenous inputs originating from the outside network. The delayed values of the outputs on which the model output $\hat{y}_N(t+1)$ is regressed are $y^{I2}{}_{Nn}(t), y^{I2}{}_{Nn}(t-1),..........y^{I2}{}_{Nn}(t-n)$. Thus the MLP RNN of **Fig. 5** corresponds to the true system model and the application in Immune system mapping in Human Antibodies clinical system is illustrated in the **Fig.6** except it now has with the repressor (past m and n inputs). The MLN RNN-NARX- Nonlinear autoregressive with external input is shown in Fig. 5 and Fig. 6 below

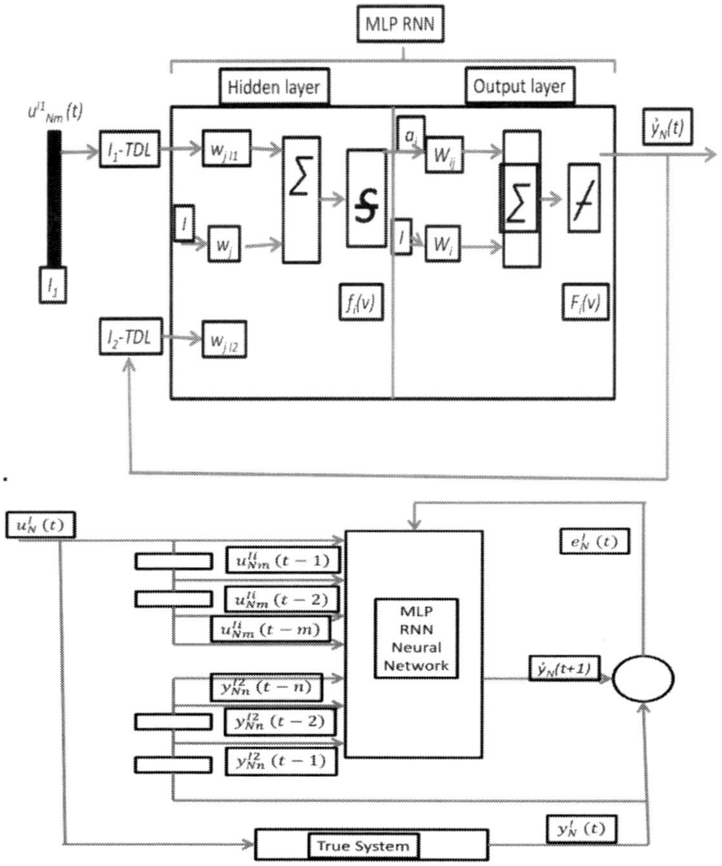

Figure 4: Decision Support System Model Prototype Architecture

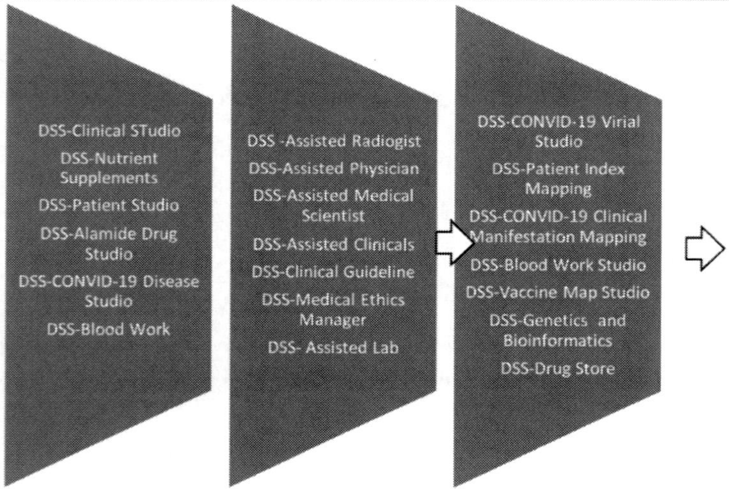

Figure 5: DSS Human CONVID-19 Immune system mapping

The need to achieve quality-assured vitro diagnosis and mapping in CONVID-19 Virus Disease (EVD] to curtail outbreak in Africa prompted the World Health Organization [WHO] to establish an Emergency Quality Assessment of In Vitro Diagnostics [IVD] for EVD. The possibility of improved procedures assisted by DSS could resolve many of the issues related to speed and accuracy of providing effective diagnosis and treatment optionsfor which incorporate enormous clinical data of preexisting evidence of safety performance, desktop review of selected clinical quality. According to report [29, 30, 31] the quality-assured in vitro diagnostics in CONVID-19 Virus Disease (EVD) outbreak in Nigeria has been established as a WHO Emergency Quality Assessment Mechanism of in Vitro Diagnostics (IVDs) for EVD. Drug therapies for CONVID-19 treatment are still under development. The clinical guidelines recommended for viral containment are 1) Early screening in symptomatic individuals using gene coding for a particular protein corresponds to a sequence of nucleotides in DNA. 2) Complete DSS supported medical data profile of patient is recommended to screen possible treatment options and to make diagnosis in borderline cases using AI technology 3) the final diagnosis must be delivered by a clinician or medical professional with expertise in genotyping and virology. The DSS diagnosis mapping is to assist with decision making process to support accurate timely and close monitoring in CONVID-19 virus containment using Artificial Intelligence (AI) technology to discriminate client cases and advice on alternative post treatment options. .

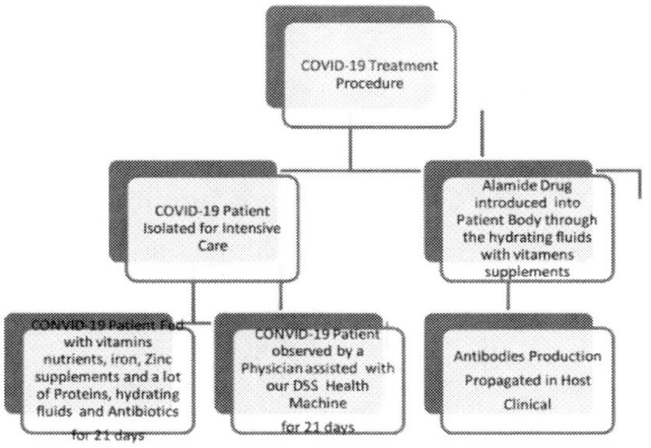

Figure 6: CONVID-19 Patient Treatment Therapy

3. RESULTS AND DISCUSSION

The following randomized clinical model simulation analysis was derived from preclinical data below and those presented in Appendix 1

3.1 Preclinical Data-No human subject was used for our research and our source of data was largely from published data, so no ethical approvals were required. The method of simulation of clinical treatment for COVID viral infected patients discussed requires no human clinical trial subject. The basis of this study is to present preclinical random simulations for a future clinical trial study after FDA approval. An extensive preclinical diagnosis of recent literature publication on Novel beta coronavirus, the 2019 novel coronavirus for the recent cluster of pneumonia cases in Wuhan China as reported Huang et.al [4] DOI: http// doi.org/10.1016/so140-6736 (20) 30 83-5 was used to understand and develop genomics human immune deficiency clinical surrogates preclinical guidelines to classify the DNA protein Identification Resource Protocols [PIR] in a Java Programme Web based PHP-Expert protein analysis bioinformatics software algorithm model. The objective of extensive literature study is to understand database structures of Nucleic Acid Sequence database Encoded in Genome Human Immune clinical expression proteomics database and search engines. The model proposed by Huang et.al DOI: http// doi.org/10.1016/so140-6736 (20) 30 83-5 required all Patients with suspected 2019-NCOV admitted to a designated hospital in Wuhan. Clinical data was prospectively collected and analyzed data laboratory confirmed 2019-nCov infection by real time RT-PCR and next generation sequencing. Data were obtained with standardized data collection forms shared by WHO and International Severe Acute respiratory and Infection consortium from electronic medical records. Patients and their families were subjected to questionnaire to derive epidemiological and symptoms data and compared with patients admitted into the intensive care unit (ICU). Data derived from the study from 41 patients admitted into the hospital were confirmed to have nCov-19 infection. Most of the infected patients were men (30(73%), including diabetes (eight (20%), hypertension (six (15%). Median age was 49.0 years (IQR 41.0-58.0), 27(66%) of 41 patients has been exposed to the seafood market. The Family cluster exhibited common symptoms at onset of illness were fever (40(98%) of 41 patients), cough (31(76%) and myalgia or fatigue (18 [44%]. The less common symptoms were sputum production (11 (28%) of 39), headache (three [8%] of 38], hemoptysis (two [5%] of 39] and diarrhea (one (3%) of 38). Dyspnea developed in 22 (55%) of 40 patients (median time from illness onset to dyspnea 8.0 days [IQR 5.0 -13.0]. 26 (63%) of 41 patients had lymphopenia. All 41 patients had pneumonia with abnormal findings on chest CT. Complications included acute respiratory disease syndrome (12 [29%]), RNAaemia (six (15%)), acute cardiac injury (five (12%) and secondary infection (four (10%)). 13 (32%) patients were admitted to an ICU and six (15%) died. Compared with non-ICU patients, ICU patients had higher plasma levels IL2, IL7, IL10, GSCF, IP10, MCPI, MIPIA and TNF. Their interpretations of their work are that 2019-nCov infection caused clusters of severe respiratory syndrome coronavirus and was associated with ICU admission high mortality rate. Figure 8 show a DSS MATLAB neural network simulator developed to provide randomized clinical results for specific patient cases based on randomized data above for CONVID-19 derived from data center Lagos hospital presented in Appendix and preclinical data above. Figure 9 shows the DSS Immune Index Mapping of patients diagnosed. Patient 3 has the highest immune index mapping followed by patient 0, 28%, patient 1, 19% and 2% for patient 3. The immune index is a measure of viral disease containing capacity inherent in the immune system and the best time to take the measurements is within the first 0-3 days when the infected cells is still healthy.

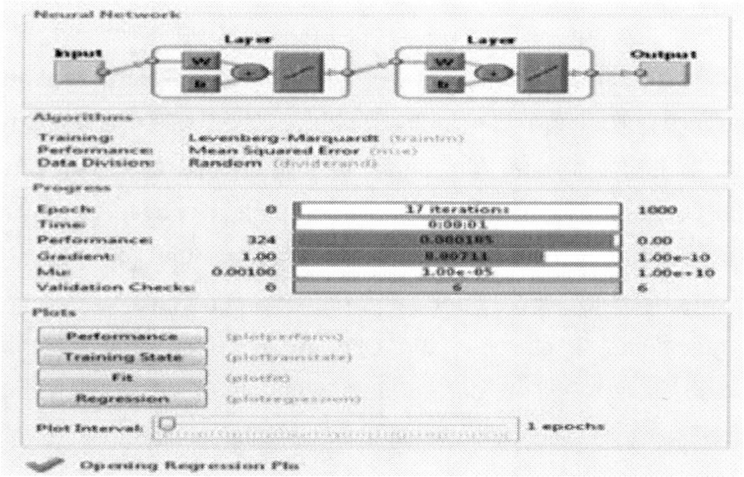

Figure 8: DSS Neural Network Simulator

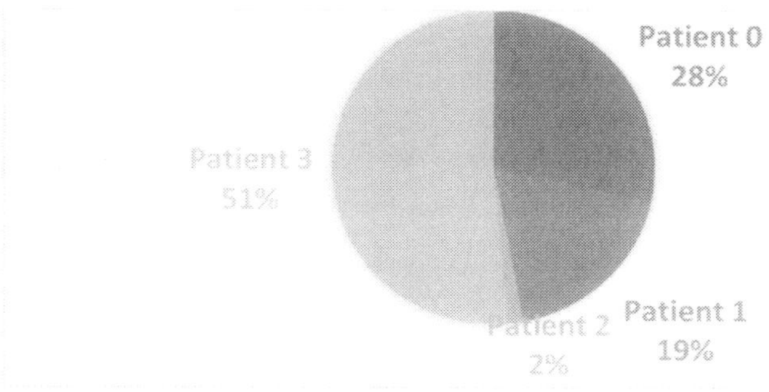

Figure 9: DSS immune index mapping

The Bayesian model is used to simulate CONVID-19 infection index and antibodies evolutions preclinical trials based on patients data are based on specific cases of hematology, genetics and patients conditions. Patient 0 shows the highest antibodies index (0.415396) within the first three days because of its higher immunity index (0.91551) compared to Patient 1, Patient 2 and Patient 3 which have immunity and antibodies potential index of (0.271027, 0.262412), (0.02204, 0.507298) and (0.751984, 0.108534) within the first one to three days. Patient 2's immunity index is considerable low (0.02204) because of its higher disease load (0.650291) compared in Patient 0, patient 1 and patient 3. Figure 10 is the cardiovascular system and support mechanism. When a human host patient is exposed to CONVID-19 virus the whole cardiovascular system is exposed as well. Patient 0 shows reduced antibodies index (0.20743) after 6-9 days because of its higher disease load of (0.776167). Patient 1 disease index have increased to (0.736847) from (0.395947) while the antibodies index (0.262412) has increased to (0.8708) while, Patient 2 which have antibodies index reduced to (0.144319) from (0.507298), and Patient 3 increase to (0.291349) from (0.108534) within the first six to nine days. Patient 2 and Patient 3 have disease load greatly reduced to (0.098344) and (0.030054) respectively.

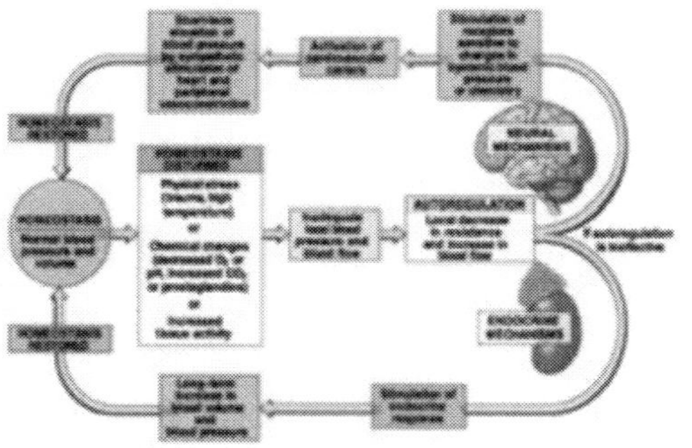

Figure 10: Cardiovascular System and support mechanism

Figure 11 shows the pie chart of DSS antibodies index of various patients. Patient 1 shows the highest antibodies index while patient 2 shows the least index. In other words, the higher the antibodies index, the higher the potential for cure of the CONVID-19 virus.

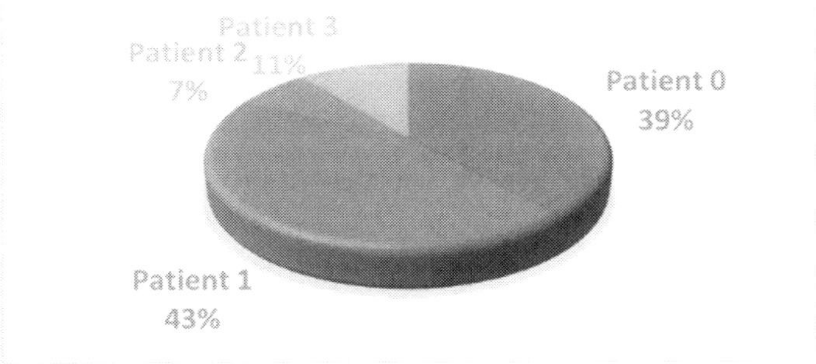

Figure 11: Piechart ofantibodies index mapping

Figure 12 shows the pie chart of various disease states of the patient. Patient 0 which is primary carrier of CONVID-19 virus exhibited the highest disease state of 47% followed by Patient 1.Patient 0 simulated diseased states of 47%, patient 1 45%, Patient 2, 6% and Patient 3%. Diseased states was modelled as probablity bayesian distribution subject to a set of diseased conditions from the degree of symptoms expereenced by each patient. Patient 3 shows the least disease states. Even though patient 1 shows some highest antibodies index requirements of 43% compared to patient 0, of 39%, the disease state in patient 0 is highest at 47% to 45% of patient 1 in comparison within 3-6days. Patient 0 shows reduced disease load to (0.401072) after 18-21 days from a high of (0.776167) within 6-9 days because of increase treatment index (0.581956) from (0.358707). The disease index for Patient 1 and Patient 2 reaches an all-time high of 0.9910997 and 0.978334 respectively resulting from a lower treatment index notwithstanding increasing treatment index from 0.107017 to 0.231775 for the remaining 18 to 21 days. Patient 2's higher disease indexes could be attributed to the lower antibodies index of 0.107017 reductions from 0.870805.

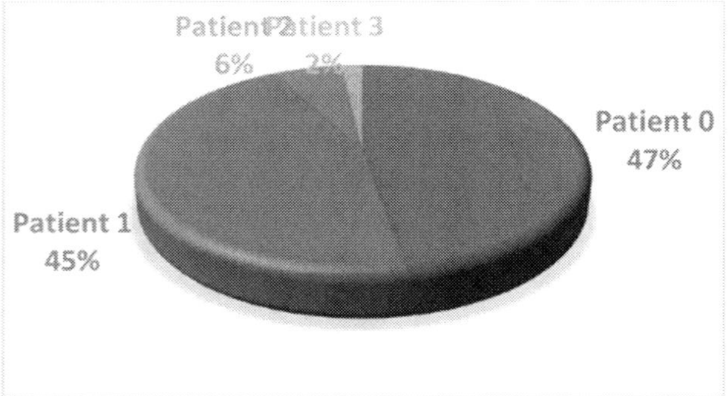

Figure 12: Patient Simulated Disease States

Figure 13 is the disease state of patient 0, with the blue (1) classification showing the state within 3-6 days and the brown (2) shows the reduction in the CONVID-19 infection to 34% within 18-21days. Figure 14 shows Patient 0 antibodies index is reduced from 62% within the 3-6days to 38% within 18-21 days, showing a reduction. The CONVID-19 virus has a lifecycle to extinction within an 18-21 days period, and if the patient can be kept alive within the 21 days, the immune system antibodies would have totally eliminated the CONVID-19 virus. So even though virus infection has no cure the findings from this work showed that the CONVID-19management of all disease symptoms within life cycle , patient immune system antibodies is functional to contain and extinct the virus within 21 days.

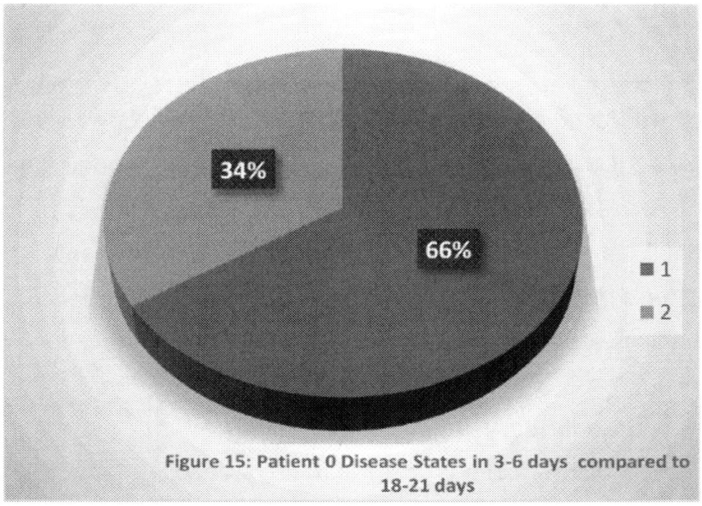

Figure 13: Patient 0 Simulated Disease states in 3-6 days compared to 18-21 days

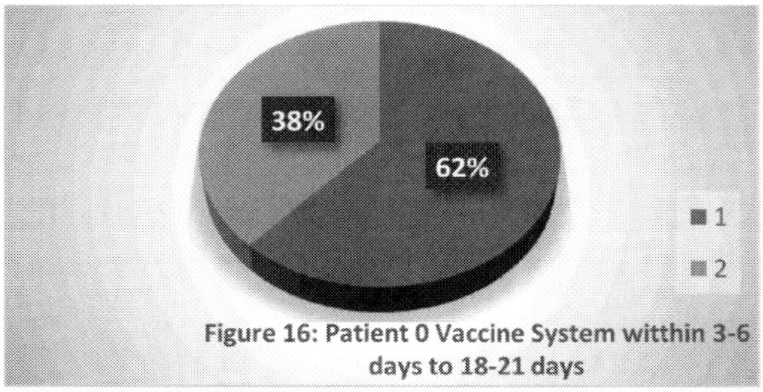

Figure 14: Patient 0 Antibodies System within 3-6 days and 18-21 days

3.2 Clinical Trials Simulation and Testing Algorithms

The key milestone and scientific discovery in this research workprogress is not whether antibodies would form in the Human Immune system, because it is natural process that must occur, but how quickly, what quality, what quantity and whether the right antibodies can be generated and the key is through nutritional supplements. The clinical software develop in our work to generate preclinical studies has been applied to find solutions to COVID-19 genomics techniques and methodology for treatment of viral infections acquired immune deficiencies (AIDS) conditions. The preclinical simulation results could support future clinical trials and planned FDA for COVID-19. Viral infections are not a disease that kills its hosts but an unstable RNA protein in the DNA proteome architecture coding system from this initial preclinical studies. So the first goal was to develop a Clinical Software DNA (RNA) neural network system based on Bayesian Distribution Models, where a variable, a diseased protein or RNA (convid-19, Viral infection) and to determine their immunity infection and antibodies indexes on a set of conditions such as nutritional supplements, DNA biomarkers, retroviral drugs. The DNA (RNA) coding was develop into MATLAB programme and Java Script by mapping the DNA 5 proteome nucleotides to several natural occurring amino acids proteins constituents (polar, non-polar or charged) by an alpnumeric coding system and covert this to binary bit coding systems and used this to generate several combinations and permutations of the various biomarkers in the genes system based on diseased symptoms from the Wuhan Publication to develop Biomarker and Nutritional supplements that can mapped Clinical Success and provide a sound basis for Clinical trials for CONVID -19 in future research work. This present research was used to develop clinical guidelines for software modeling and simulation of possible outcomes for CONVID-19. The preclinical guidelines and model algorithms submitted in DSS modeling for CONVID-19 assessment and treatment is the outcomes of the results evidence derived from data in Lagos, Nigeria. This success should inform the premise of a more careful clinical study of the clinical cases as reference for a global application based on science and model simulation algorithms and clinical guidelines presented in this research work.

3.3 CONVID-19 Virus Patient Index, Primary and Secondary Newborn Screening

Classes in Clinical Guideline- 1.1 Summary of recommendations 2.1 Methods of development of evidence base recommendations 3.1 Definition, path physiology, mode of transmission, evaluation, and morbidity 4.1 Screening, evaluation and clinical testing algorithms (Biochemistry, microbiology, bioinformatics and genetic reengineering, epidemiology of CONVID-19 virus infection, clinical testing algorithms, bio electrochemistry, immunology responses and deficiency)

3.3.1 Prenatal Treatment of CONVID-19 Virus-Diagnosis of CONVID-19 Virus disease condition following infection and transmission by Index 1.2 medical treatment of CONVID-19 virus in index, primary and secondary patient (antibodies models, testing algorithms, complications and side effects, -clinical trials, experimental therapy)

3.3.2 Research Methodology, Materials and MethodsSample Population, Study design, laboratory methods, microbiology, biochemistry, alamide pro drug design synthesis, radiological engineering, MRI CT Scan, patient zero antibodies synthesis, extraction and purification, soxhlet extraction, testing kits, analysis/ clinical trials, CONVID-19 virus in adulthood and mental health

3.3.3 Prognostic Markers Assessments *(Ethical concerns,* data collation methods and statistical analysis, results, evaluation of CONVID-19 virus testing algorithms and clinical trials, stability of screening)

3.4 CONVID-19 Intervention Program (Appendix, Reference, Order Form)

Transient blood flow of several stages of cerebral aneurysm simulated over a period of 1.00 seconds at zero by increasing disease factor k_D was simulated in a MATLAB program that uses method of characteristics. The following example shows you how to solve a parabolic equation in stages and how to set an initial condition as a variable: At the MATLAB command prompt, type pde tool. Figure 14 is viral load pressure distributionindex along z, r and time axis

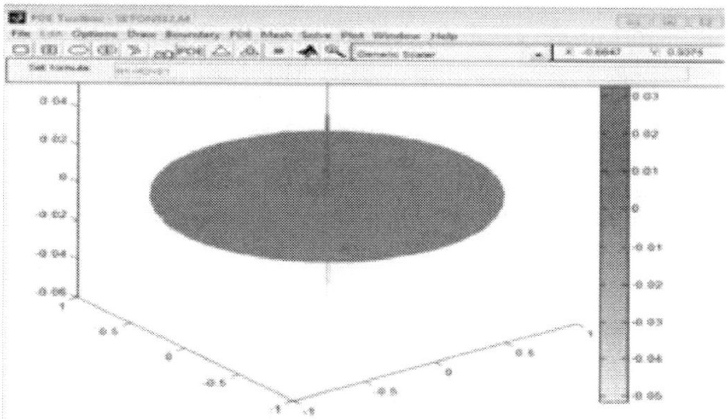

Figure 14: PDE tool box displaying a plot of the initial boundary conditions.

The MAT file derived from the PDE tool is feed into the neural networks tool box as a source of raw data in other for the fitting networks, Perceptions and predictive control networks to be trained. In this work, it was saved as data.mat and for the curve fitting networks. Figure 15 is the neural networks data manager showing the functions and data imported from the model in the PDE tool. All variables represented by letters in the network data manager signify various vectors and matrices derived from the CONVID-19viral loadmodelthat have been further simplified with simplified data presented in Appendix

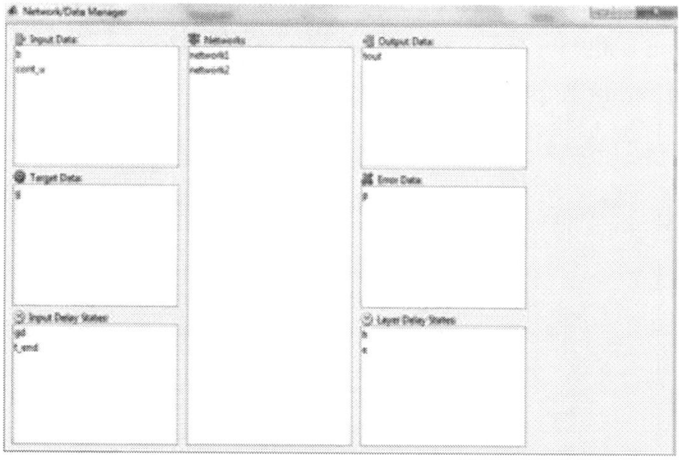

Figure 15: MATLAB neural networks data manager

In current study main aspects of the clinical software and randomized modeling are addressed to 1) provide scientific basis based on precedents data and publications to progress viral research in CONVID-19. All viral infection is a single RNA affecting specific biomarker. The current research highlights antibody biomarker affected and deficient in COVID-19 cases. 2) To establish all viral infection, whether COVID-19 as having the same immunity and antibody index except different virus affect different biomarkers. COVID-19 affects the biomarker in corona tract and means of transmission is airborne unlike Ebola virus propagated through blood distribution system and HIV affects CD4+ markers, and all these are acquired human deficiency virus. The studies in the clinical software could also be applied non-acquired are Cancer, Cerebral aneurysm and Alzheimer's, Type 2, 3 diabetes on input of the relevant data. All viral infections have no drug generic cure. The cure can only happen through antibodies that shorten their lifecycles or restore stability through our duality theorem and laws of energy/matter creation and destruction 3) providing the science and technology methodology using antibodies to contain CONVID-19 viral infection based on certain nutritional supplements. Figure 16 shows the MATLAB neural network GUI displaying external and internal neurons.

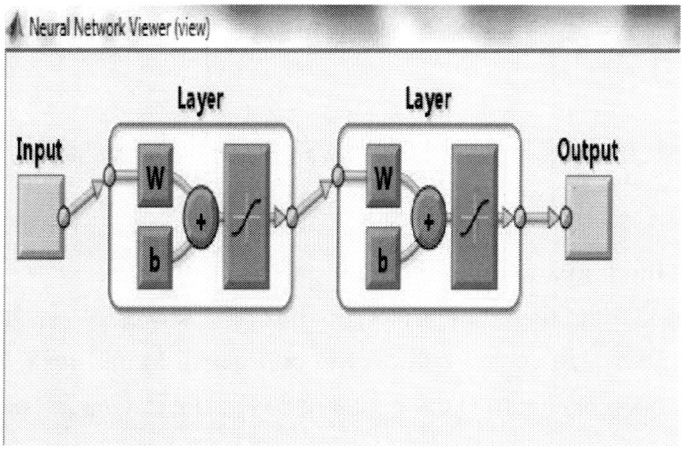

Figure 16: MATLAB neural network viewer

Figure 17 shows the neural network data fitting tools used to simulate the neural network input-output systems and implement the regression neural network algorithms. Figure 17 is the general graphic user interface (GUI) of neural network training tool and a link for the outcomes of the training like the performance, training state, fit, and regression. Here is the performance plot for the performance.

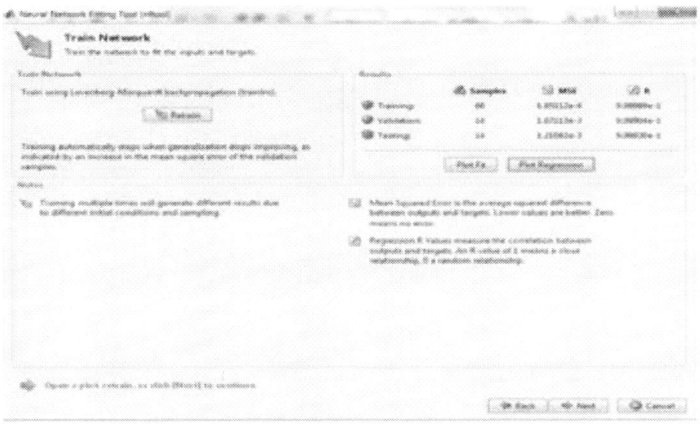

Figure 16: MATLAB neural network fitting tool

The network is to be trained using levenberg-Marquardt back propagation. Levenberg-Marquardt is the best training algorithm adopted for complex problems and the application of multiple networks in solving problems like CONVID-19 epidemiology production in human immune systems

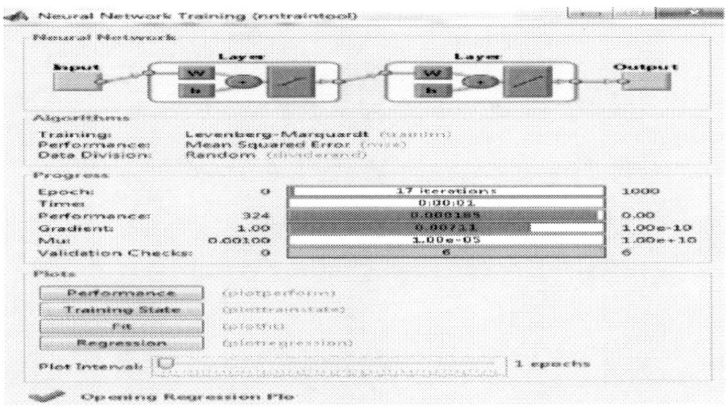

Figure 17: MATLAB neural network training tool

Figure 18 shows performance plots for 11 iterations epochs for CONVID-19 immune –epidemiology iterations derived from data on the viral infection of Ebola Epidemic. This training stopped when the validation error increased for six iterations, which occurred at iteration 23. If you click Performance in the training window, a plot of the training errors, validation errors, and test errors appears, as shown in the following figure. In this example, the result is reasonable because of the following considerations: The final mean-square error is small. The test set error and the validations set error have similar characteristics. No significant over fitting has occurred by iteration 17 (where the best validation performance occurs)

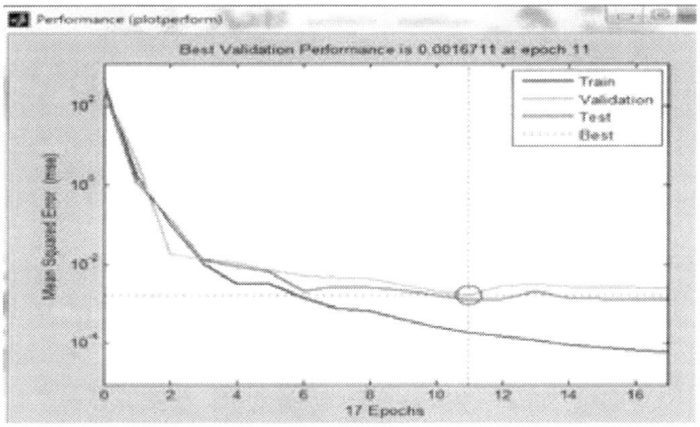

Figure 19: **Performance plot**

Figure 20 shows CONVID-19 simulation validation charts of DSS in the training state. These plots are characteristic to a group of CONVID-19 infection distributions of viral vectors from Ebola epidemic in 2014 fed from the model to be used in predictive control and decision support about the virus epidemiology levels. Any deviation from this plot will be recorded and seen as anomaly forCONVID-19 disease manifestation.

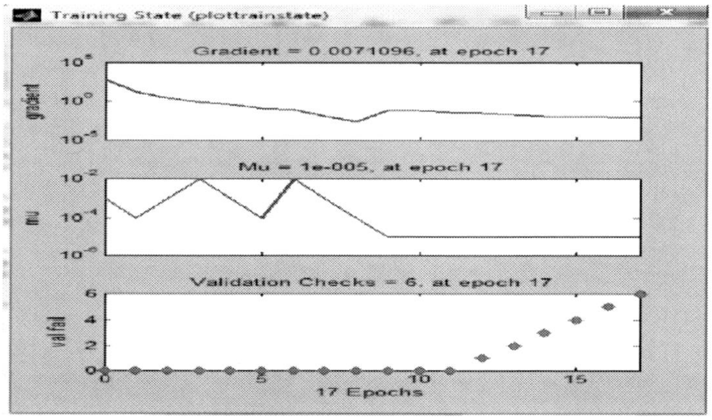

Figure 20: DSS Training state

Figure 21 is a plot fit between the clinical data input functions (the CONVID infection distribution data during CONVID-19infection –immune antibodies simulation responses and the clinical output of data which is the disease factor C_f and the gene protein targets which is the true value the dimensionless pressure used in the IMPES PDE model. The training targets are shown by the blue circles while training inputs are the blue +, validation targets and Output d is green respectively while test details are red, Errors appear orange the fit is black.

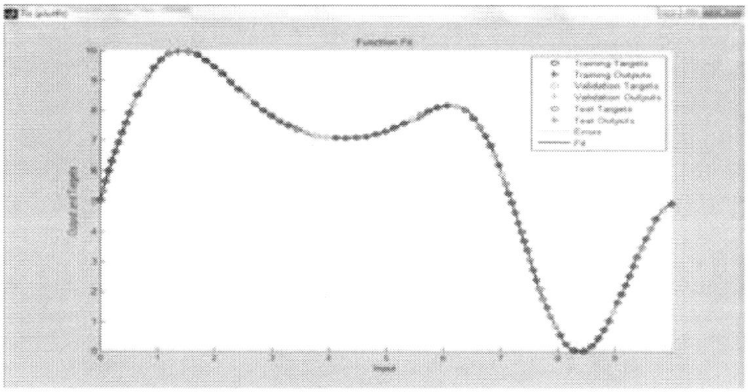

Figure 21. DSS Neural Network Fit

Figure 22 is the regression analysis for the viral clinical data load network trained. This model feature predict future rate of disease and in the decision making process of a clinical team. All the output from the neural networks can be used to design anembedded software system that has been programmed to generate these outputs when fed with data from the production system. This system generates output, stores them in its memory and compares them with the previous output generated.

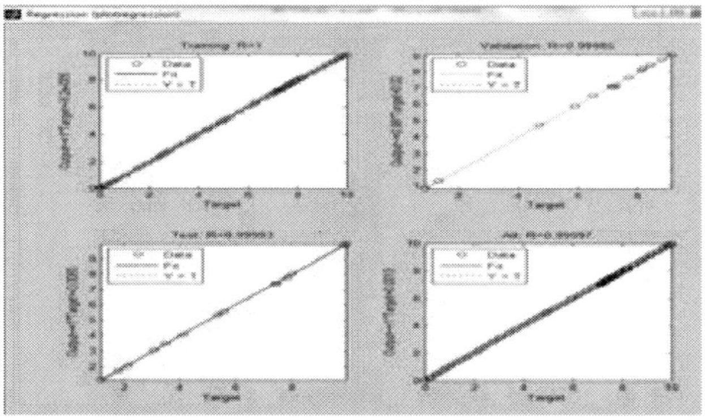

Figure 22 DSS Regression plot

A transition from a point of minimal disease to a critical disease can be noticed by this system. DSS would trigger an alarm or other forms of communications (probably red light) to inform the clinician personnel of a critical CONVID-19 disease transition.

3.5 CONVID-19 infection Pathology Blood Pressure Variation along the Vessel Length

Two important discussions of our study relate to specific anatomical and physiologic parameters to post-treatment fluid dynamics into CONVID-19 infection pathology. Figure 23, Fig 24 and Fig 25 shows the time evolution of the internal blood pressure at various k_D values (0, 1 and 5) to show pathological trends of in CONVID-19 infection lifecycle growth predicted by DSS model on a clinical basis along the length of a disease artery. There are several ways that quantitative knowledge of the relationship between coil packing density and aneurismal inflow could facilitate improved outcomes for patients treated with coil embolization. At each disease factor considered, the blood pressure was observed to generally steady out to about at the outflow

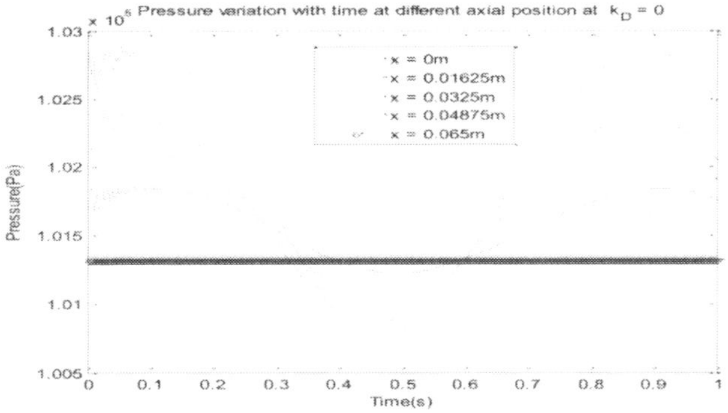

Figure 23: DSS Numerical Pressure Variation with Time at Different Axial Position k_D =0

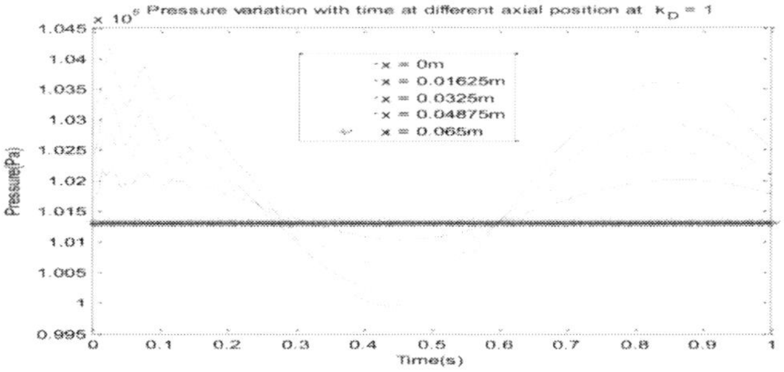

Figure 24: DSSS Numerical Pressure Variation with Time at Different Axial Position k_D =1

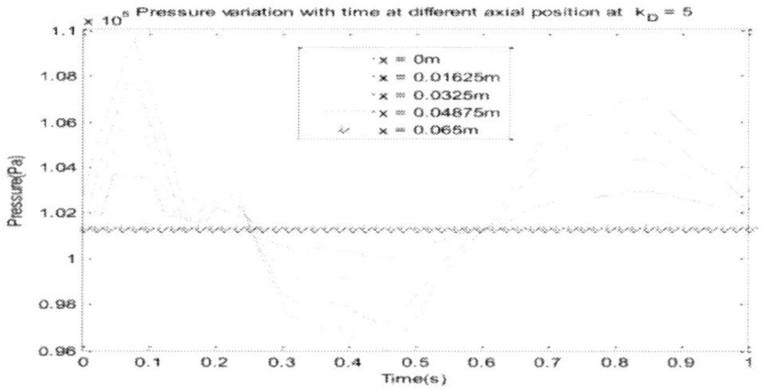

Figure 25. DSS numerical pressure Variation with Time at Different Axial Position k_D =5. This distinction in flow rate along the vessel becomes more pronounced as the disease factor increases as disease factor in CONVID-19 infection lifecycle.

4.0 CONCLUSION

The CONVID-19 clinical manifestation was simulated as possible postulates for providing scientific basis the science of human DNA genomics immune organization, bioinformatics evolution and antibodies therapy of coronavirus (CONVID-19) epidemiology. The CONVID-19 Biomarker in human immune physiology pathology targets single-nucleotide polymorphisms derived from phenome genetics of the patient physiology and disease symptoms. The science was further developed into mathematical Programme in MATLAB software environment to demonstrate virus infection epidemiology and antibodies mapping algorithms and clinical simulation of CONVID-19 epidemiology. A fit of clinical expectations could be observed for the infection plots. Decision support systems has been developed to evolve clinical guidelines specifications to assist clinicians in making decisions related to CONVID-19 virus containment, evaluation of antibodies models, therapeutic treatment methods and development of clinical testing algorithms in Lagos, Nigeria. The results of simulation showed the reduction in the CONVID-19 infection to 34% within 18-21days. The Patient 0 Antibodies System is reduced from 62% within the 3-6days to 38% within 18-21 days. This reduction in the CONVID-19 virus has a lifecycle to extinction within 18-21 days period. Further recommendation to the clinician assumes within the 21 days, the patient immune system will contain the CONVID-19 virus under it eventually extinct. DSS was developed to enable the clinician make critical decisions to the patients' health and to provide solutions to the CONVID-19 epidemics by advising clinicians and public health therapy on alternative treatment to the CONVID-19 infection. The preclinical clinical model data and algorithms inputted in DSS modeling of CONVID-19 assessment and treatment is the based outcomes of the results from clinical data obtained in Lagos, Nigeria. Two important discussions of this research study relate to specific anatomical and physiologic parameters to post-treatment for CONVID-19 infection pathology.

ACKNOWLEDGEMENTS

The author acknowledges the support of University Technology System Limited, Syntechsys Corporation for support of the research work. Also the following colleagues and students of Department of Chemical Engineering are acknowledged for their contribution to work. The support of the Center of Disease Control (CDC) data center and the Lagos University Teaching hospital in Nigeria for access to data used in the simulation. The head of Chemical Engineering and for all staff of Unilag is duly acknowledged. Dr. Kingsley E. Abhulimen also acknowledges the support of the University Technology System Limited of Syntechsys Global Network Inc.The contributions of Edwin Keusey of Keusey and Associates, P.C for suggestions of useful information to development of the paper are appreciated. Also the contributions med studio simulator design specifications are appreciated.

DECLARATION OF INTERESTS

Author declares no competing interests.

Main Contributions of this work to the science community;

1. Providing scientific basis based on precedents data and publications to progress viral research in CONVID-19. All viral infection is a single RNA affecting specific biomarker. Our research highlight which biomarker has been affected and deficient in COVID-19 cases.
2. To establish all viral infection, whether COVID-19 as having the same immunity and antibody index except different virus affect different biomarkers. COVID-19 affects the biomarker in corona tract and means of transmission is airborne unlike Ebola is through blood distribution system and HIV affects CD4+

markers , and all these are acquired human deficiency virus. The non acquired are Cancer, Cerebral aneurysm and Alzheimer's, Type 2,3 diabetes. All viral infections have no drug generic cure. The cure can only happen through antibodies that shorten their lifecycles or restore stability through our duality theorem and laws of energy/matter creation and destruction

3. Providing the Science and Technology methodology using antibodies to contain CONVID-19 viral infection based on certain nutritional supplements
4. Providing basis for a clinical trials and our research would serve as the Science of why CONVID-19 can be contained based on precedent success with application to published work on virus epidemic in West Africa

NOMENCLATURE

1-Means Very Good

Patient 0-Primary Carrier Patient 1-Secondary Contact to primary Carrier (Male)

Patient 2—Secondary Carrier to Primary Carrier but different sex (Female).

Patient 3-Tertitiary Carrier

$\theta_{Nm}^{I_i}(t)$for past inputs and $\theta^{I_2}{}_{Nm}(t)$ for past outputs;

$g^i{}_N(-)$, is the unidentified nonlinear function stimulated by the MLP RNN;

θ_N^i; is an unknown vector of adjustable parameters of a network (i.e., the joint weights);

$e^i{}_N(t)$is white noise independent of the inputs.

y=output

u=input

W=weight

b=bias

N=data set

w_{11} w_{12} ...w_{1R} Synaptic weights

R Number of elements in input vector

S Number of elements in input vector

$u^L{}_N(t)$ Input signals

$y^L{}_N(t)$ Observed Outputs

$y^i{}_N(t)$ Corrected Predicted Output

$\phi^{L_{1,2}}{}_{Nm}$ Regressor Vector

$\phi^{L_1}{}_{Nm}(t)$ Pass inputs

$\phi^{L_2}{}_{Nm}(t)$ Pass outputs

$g^i{}_N(-)$ Unidentified

$\phi^i{}_N$ Non linear function, unknown Vector adjustable parameter

W_{ij1} *and* W_{ij} are hidden and output weights

w_j *and* w_j are hidden and output bias

$J_i = TSE$ (Total Square Error)

REFERENCES

1. Na Zhu, Dingyu Zhang, Wenling Wang, Xingwang Li " A novel Coronavirus Froma Patients with Pneumonia in China, 2019, NEng J Med 2020; 382:727-733 doi: 10.1056/NEJ Moa 2001017
2. Su S, Wong G, Shi W " Epidemiology, Genetic recombination, and Pathogenesis of corona virus Trends Microbiol 2016; 24: 490-502
3. New york Times Published March 2020
4. Chaolin Huang, Yeming Wang, Xingwang Li, LiliRen, Jianping Zhao, Yi Hu, Li Zhang, Guohui Fan, JiuyangXu, XiaoyingGu, Zhenshun Cheng, Ting Yu, Jiaan Xia, Yuan Wei, Wenjuan Wu, ueleiXie, Wen Yin, Hui, Li, Min Liu, YAN Xiao, Hong Gao, Liguo,JungangXie, Guangfa Wang, Rongmeng Jiang, ZhanchengGao, Qi Jin, Jianwei Wang, Bin Cao "Clinical Features of Patients with219 Novel coronavirus in Wuhan, China" Lancet Vol 395, issue 10223, p497-506, 2020
5. Weiss SR, Leibowitz JL. CornavirusPathogenesis .Adv Virus Research 2011, 81-85 -165Holmasen, I. (1990).
6. Altschul, S.F., Madden, T.L Scaffer, A.A., Zhang, J., Zhang, Z., Miller, W. and Lipman , D.J (1997) Gapped BLAST and PSI-BLAST: a new generation of proetein database search programmes , Nucleai Acids. Res., 25, 3389-3402
7. CDC Centers for Disease Control and Prevention (www.CDC.gov) Coronavirus (CONVID-19) 2020
8. Wheeler, D.L, Church D.M., Federhen, S., Lash A.E., Madden T.L., Pontiu, J.U., Schuler, G.D, Schriml, L.M., Sequera, E., Tatusova and Wagner L. (2003) , Database resources of the National Center for Biotechnology, Nucleai Acids Res. 31, 28-33
9. Stein, L (2001), Genome Annotation: From Sequence to Biology, Nature Reviews (Genetics), 2, 493-503
10. Branden, C.I and Tooze, J. Introduction to Protein Structure (2nd Edn. New York; Garland, 1999
11. Lesk, A.M. Introduction to Protein Science: Architecture, Function and Genomics (Oxford: Oxford University Press, 2004
12. Arthur M. Lesk "Introduction to Bioinformatics, 2nd. Ed Genomics (Oxford: Oxford University Press, 2000

13. Kaslow Francis " The Epidemiology of AIDS" Newyork Oxford, University Press 1989
14. Gallo RC, Wong-Staal F, Montagnier L, Haseltine W, Yoshida M (1988), Nature 333:504g
15. Chiu I.M, Yaniv A, DahlberJE, Gazit A, Skuntz SF, Tronick SR, Aaronson SA (1985): Nature 316:366
16. Arya SK, Guo C, Josephs SF, Wong-Staal F (1985): Science 229:69
17. Sodroski J, Patarca R, Rosen C, Wong-Staal F, Haseltine W (1985): Science 229:74
18. Genomics (Oxford: Oxford University Press, 2004
19. Weekly epidiological record, 49: No.41, pp 341-343 (1974)
20. Abhulimen KE. Theoretical model basis of clinical guidelines for CONVID-19 viral infection containment in Lagos, Nigeria. Virol Res J. 2017;1(2):16-42.
21. United States Department of Health, Education and Welfare; Center for Disease Control, Isolation Techniques for use in hospitals, Washington, DC, US Government Printing Office, 2nd ed. 1975 (DHEW publication No. (CDC) 76-8314)
22. Madeley C.R. Guide to collection and Transport of viriological specimens, Geneva, World Health organization, 1211 Geneva 27, Switzerland (telegraphic address: Unisante Geneva)
23. Banana. *In the New Encyclopedia.* Vol.1. Chicago. Encyclopedia Britannica, Inc. Pp. 855-856.Food and Agriculture Organisation of the United Nations (FAO). (2003). Banana Fruits and their Uses. Retreived June, 2010, from FAOSTAT Statistics Database.
24. Frezier, C.W. and Westhoff, C.D. (1995). *Food Microbiology. 4th Edition.* New Delhi. Tata McGraw-Hill. Pp 339
25. Agrawal, A.K and Pradeep, P. (2006). *Industrial Microbiology: Fundamental and Application.* U.S.A. AVI Publishing Company, Inc. Pp 451-458.
26. Akubor, P.I., Obio, S.O., Nwadomere, K.A. and Obimah, E. (2003). "Production and Quality Evaluation of Banana prodrug solution". *Journal of Plant Foods for Human Nutrition.* **58(3),** 1-2.
27. Sistrom CL, Dang PA, Weiburg JB, Dreyer KJ, Rosenhal DI, Thrall JH. "Effect of Computerized order entry with integrated decision support on the growth of outpatient" procedure volumes: seven-year time series analysis. Radiology 2009, 51: 14
28. Stivaros SM, Sinclair D, Bromiley PA, Kim J, Thorne J, Jackson A. Endoscopic third ventriculostomy: predicting outcome with phase contrast MR imaging. Radiology 2009: 252:
29. Abhulimen K.E., "Modeling Decision Support Systems (DSS) FOR OPTIMAL diagnosis and Treatment of Cerebral Aneurysm" ASME Journal of Engineering and Science in Medical Diagnostics and Therapy JESMDT-18-1021 (2018)
30. Weekly epidiological record, 49: No.41, 1974, pp 341-343
31. United States Department of Health, Education and Welfare; Center for Disease Control, Isolation Techniques for use in hospitals, Washington, DC, US Government Printing Office, 2nd ed. 1975 (DHEW publication No. (CDC) 76-8314)

32. Madeley C.R. Guide to collection and Transport of viriological specimens, Geneva, World Health organization, 1211 Geneva 27, Switzerland (telegraphic address: Unisante Geneva)

Enclosed are copies of the following patent documents that we consider relevant to our research study
Patent Document No. Inventor
US 2005/0080260 A1 Mills et al.
US 2005/0220772 A1 Chow et al.
US 2008/0003183 A1 Guo
US 2008/0113925 A1 Harley et al.
5,587,463 Sessler et al.
5,776,966 North
US 6,455,555 B1 Nakagawa et al.
US 6,653,130 B2 Rios
Foreign Patent Document No. Country & Date
WO2005044365 A2 WIPO; 5/2005
CN1948475 A China; 4/2007
CA2631714 A1 Canada; 5/2008
GB2446310 A United Kingdom; 8/2008

"Can a Bone-Marrow Transplant Halt HIV?";
http://www.time.com/time/health/article/0, 8599, 1858843,00.html; printed on

APPENDIX 1

Table A1: Clinical Data from a Clinic in Lagos, Nigeria to probe viral infection, malaria parasites and cerebral deficiencies *[Ethical Methods: Names of Patient removed and replace with Index and Clinical Settings: Hospitals. Clinics and Locations Referenced are removed and replaced with Clinical Settings Number to Protect Sources FBC. MP.PT. Widal represents Patient Test Parameters]*

S/N	CLINICAL SETTINGS	Patient Index	Patient Age	Patient Sex	Patient Test Parameters
1	1269	J1	Ad	F	FBC, MP, PT
2	1270	J2	1 Month	M	FBC, ,MP
3	1271	J3	Ad	F	FBC,MP, widal
4	1272	J4	2yr	M	FBC, MP
5	1273	J5	Ad	M	MP,Widal
6	1274	J6	4yr	M	FBC,MP

7	1275	J7	2yr	M	FBC,MP
8	1276	J8	3yr	F	FBC,MP,Widal
9	1277	J9	8yr	M	FBC,MP
10	1278	J10	19yr	F	FBC,MP

Table A2: Patient Blood Work (Hb, hemoglobin, PCU (Packed Cell volume, Total WBC))

[Ethical Methods: Names of Patient removed and replace with Index and Clinical Settings: Hospitals. Clinics and Locations Referenced are removed and replaced with Clinical Settings Number to Protect Sources FBC. MP.PT. Widal represent Patient Test Parameters] Data published

S/N	CLINICAL SETTINGS	Patient Index	Hb(hemoglobin) Concentration	PCU(Packed Cell Volume)	Total WBC(White Blood Cell Content)
			X1	**X2**	**X3**
1	1269	J1	12.8	39	4.3
2	1270	J2	13.4	40	7.1
3	1271	J3	11.7	34	5.6
4	1272	J4	13.1	41	5.7
5	1273	J5			
6	1274	J6	10.9	33	5.7
7	1275	J7	11.4	34	7.9
8	1276	J8	14.7	43	17.9
9	1277	J9	6.8	33	79
10	1278	J10	9.3	3.8	6

601270 High Throughput Process Development for CEX Chromatography

Abstract:

Chromatographic operations are the workhorse for protein purification in the biopharmaceutical industry due to their unparalleled separation efficiency and capacity. Typically, chromatographic processes consist of multiple polishing steps, at least one of which is operated in a bind and elute mode. Although bind and elute chromatography is very effective at removing product variants, it suffers from limited capacity and long processing times. In this work, we developed efficient cation exchange chromatography (CEX) flowthrough steps for mAbs using a combination of high throughput screens and lab scale experiments. CEX resins were screened across parameter space using high throughput screening techniques. Conditions which improved process performance as compared to a bind and elute operation were identified. Following high throughput screening, this process was evaluated at the lab scale for further development and refinement. Comparable product quality and process consistency between the Tecan and the lab scale experiments suggest an efficient and reliable high-throughput process development workflow for CEX flowthrough operations.

Copyright © American Institute of Chemical Engineers. All rights reserved.

605951 The Use of Tangential Flow Filtration Systems in Synthetic Peptide Purification

Abstract:

Tangential flow filtration (TFF) is a pressure and shear driven separation process that uses membranes to separate components in a liquid solution based primarily on size differences. TFF processes such as microfiltration and ultrafiltration are widely used in the pharmaceutical industry for concentration and buffer exchange in biologics. Due to the nature of TFF membrane materials, these unit operations have been primarily limited to aqueous or low-solvent processes. Example TFF applications include monoclonal antibodies and microbially derived proteins and peptides.

Recent advances in membrane technology have enabled the fabrication of membranes with smaller pore sizes and composed of materials with improved compatibility to organic solvents. For larger and more complex synthetic peptides, ultrafiltration may provide a superior alternative to distillation and/or chromatography for concentration and solvent exchange.

This presentation will highlight the use of ultrafiltration as part of a synthetic peptide process to concentrate purified peptide solution and to perform a solvent exchange in preparation for precipitation of the pure API. This work represents a novel transfer of TFF technology from the biologic/microbial platform to the synthetic peptide purification platform.

Copyright © American Institute of Chemical Engineers. All rights reserved.

608933 In-Situ Monitoring during Mixing for Protein Crystallisation

Abstract:

Bio-active peptides are the next generation of pharmaceuticals which lead to a significant increase in demand over the last years [1]. Remarkable improvements in the upstream processing of peptides has been achieved, leading to a shift in the bottleneck to downstream purification [2]. Recently, crystallisation has emerged as a novel technology to replace expensive downstream purification techniques in the production line of bio-active peptides in order to increase economic efficiency [3, 4]. Due to their high molecular weight, low diffusivity, and soft and fragile nature, peptides require a gentle, but intensive mixing in order to achieve the same level of crystallisation reproducibility at larger scales than microlitres. The lack of suitable mixing strategies for large-scale peptide crystallisation, while guaranteeing high crystallisation throughput and reproducibility of the crystal qualities, has been identified as one of the main limitations that has to be overcome in order to apply crystallisation as a sustainable downstream purification technique

In this study, the importance of mixing on the controllability and reproducibility of the crystallisation process in large-scale batch crystallisers will be discussed. Instead of commonly used mixing strategies, such as over-head stirring, that usually leads to a more uncontrolled crystallisation with a poor crystallisation reproducibility, the impact of a gentler mixing strategy by using mechanical, oscillatory shaking will be reported. Furthermore, a newly developed image-analysis tool specific to peptide crystals will be presented. It is used to determine crystal sizes and distribution in-situ by processing images of crystals to evaluate the impact of mixing. A simple model peptide with a low solubility and molecular weight, which is known to crystallise, is used to investigate the importance of mechanical mixing on achieving a reproducible peptide crystallisation process. A tubular shaped batch crystalliser with volumes from the millilitres to the tens of millilitres along with different geometries was utilised. The crystallisation condition, such as pH, concentration, shaker speed, templating additives, and the orientation of the crystalliser to the shaking plane have been optimised. The impact on the crystallisation process is evaluated by means of measuring induction time, crystal yield, crystal sizes, and size distribution.

The induction time was determined from peptide concentration by UV-vis spectroscopy every 20-30 minutes and compared to optical microscopy images. These images were analysed using a novel automated evaluation-routine we developed to evaluate the crystallisation behaviour and crystal properties. This new computer-based image evaluation-routine enables the autonomous processing of hundreds of peptide crystal images. Thereby, commonly applied image processing steps, such as background noise filtering and image-gradient based edge determination for obtaining the crystal contours, have been utilised to determine the macro-scale crystal properties, such as crystal size distribution. This novel evaluation-routine is less expensive, able to calculate the macro-scale crystal properties within a couple of minutes, and only requires tens of microlitres of the crystal suspension instead of tens of milligrams of solid crystals, which is required in commonly used analytical techniques such as dynamic light scattering. Furthermore, it enables the evaluation of hundreds of crystals with sizes from the micrometre to the tens of centimetres scale. This makes the evaluation-routine suitable for studying macro-scale crystal properties for expensive and poorly soluble peptides in-situ, allowing to optimise crystallisation parameters during the crystallisation process. Comparison of crystal size distribution obtained with the new routine and manually obtained crystal sizes was used to optimise and identify problems within the image processing steps.

Monitoring the peptide's concentration gradient in crystallisers with volumes of up to 15 millilitres found that an increase in oscillatory shaking speed from 50 to 150 rpm leads to a reproduceable higher homogeneity of the peptide concentration within the crystalliser resulting in a higher consistency of the crystallisation processes. Additionally, higher shaking speed lead to a faster decrease in peptide concentration implying a reduce in induction time and increase in crystal growth rate. These results can be supported by investigation of the crystal size distribution, obtained with the evaluation-routine, showing that a shaking speed of 150 rpm results in a more uniform crystal size distribution with a smaller span whereas a shaking speed of 50 rpm results in a wider spread size distribution. Furthermore, the reproducibility of controlling nucleation with templating additives, such as amino acids, can be enhanced significantly when a faster shaking speed is utilised, leading to a more homogeneous distribution of the additives. Mounting the tubular crystalliser in different orientations to the mixing plane found that if the axial direction of the crystalliser is aligned to the mixing plane, higher mixing length scales can be achieved resulting in an increase of mixing efficiency while keeping the energy input constant.

This study demonstrates the importance of mixing to achieve a high reproducibility of the crystallisation process in crystalliser volumes up to the tens of millilitres for peptides with low solubilities and molecular weights. Higher shaker speed enhances the mass transfer leading to a faster crystallisation process and a narrower crystal size distribution. To study the impact of mixing on the macro-scale crystal properties in-situ, hundreds of crystal images in combination with a novel developed evaluation-routine were evaluated.

Bibliography

1. Kastin, A.J., *Handbook of Biologically Active Peptides*. 2 ed. 2013: Academic Press. 2032.
2. Guillemot-Potelle, C., et al., *Cost of Goods Modeling and Quality by Design for Developing Cost-Effective Processes.* BioPharm International, 2010. 23(6): p. 26-35.
3. Yang, H., et al., *Development and Workflow of a Continuous Protein Crystallization Process: A Case of Lysozyme.* Crystal Growth & Design, 2019. 19(2): p. 983-991.
4. Roque, A.C.A., et al., *Anything but Conventional Chromatography Approaches in Bioseparation.*Biotechnology Journal, 2020.

Copyright © American Institute of Chemical Engineers. All rights reserved.

609025 The Use of Nanofiltration Systems in Synthetic Peptide Processes

Abstract:

Tangential flow filtration (TFF) process such as microfiltration and ultrafiltration has been widely used in the pharmaceutical industry for biologics to remove smaller molecular weight process waste from the proteins. Due to the nature of these membranes, this unit operation has been limited to aqueous systems and larger molecular weight assets. Recent advances in membrane technology have enabled the fabrication of membranes with significantly smaller pore sizes and composed of materials with markedly improved compatibility to organic solvents. These changes make TFF systems a useful alternative to precipitation, distillation and solvent exchange operations in synthetic chemistry steps in small molecules and peptide processes, and in particular for amorphous or thermally labile materials.

This presentation will highlight the use of nanofiltration as part of a synthetic peptide process to remove small molecular weight by-products, to concentrate reaction solution at low temperatures and to perform a solvent exchange between chemical transformation steps.

Copyright © American Institute of Chemical Engineers. All rights reserved.

614024 (Invited Talk) Developing Antibody-Based Nanoplasmonic Barcode Biosensors for Sars-Cov-2 Detection

Abstract:

The emergence of COVID-19 as an international pandemic has caused unprecedented disruptions to modern life. As economies reopen and society adjusts to a new normal, there is a critical need to not just diagnose currently or previously infected individuals but also to be able to rapidly detect the presence of viral particles in the environment. As supported by an NSF RAPID grant, this talk will describe work the authors are doing to develop antibody-based sensor devices to meet this need. Three strategies are being pursued in parallel to identify viable pairs of antibodies: *in silico* design of novel binders, rational engineering of SARS-CoV binding antibodies, and *in vitro* screening of antibodies from the spleens of MERS-immunized mice. Once candidate antibodies are identified, they are subsequently assessed for their ability to detect SARS-CoV-2 in complex fluids via a sandwich assay in a nanoplasmonic barcode device.

Copyright © American Institute of Chemical Engineers. All rights reserved.

614095 (Invited Talk) Development of Lateral Flow Tests for Sars-Cov-2 Antibody and Antigen Detection

Abstract:

As SARS-CoV-2 began to spread across the United States, it became clear that rapid and reliable tests for antigen and antibody were needed. The mission at NOWDiagnostics is to develop simple, accurate, and easy to administer diagnostic tests that yield results in minutes. In response to the pandemic, we quickly shifted focus to the development of antibody serology tests, antibody saliva tests, and antigen saliva tests using the existing AdexusDx platform. Here I will discuss some of the obstacles we faced, as well as the successes.

Copyright © American Institute of Chemical Engineers. All rights reserved.

AUTHOR INDEX